The New Science of Breath

Coherent Breathing for Autonomic Nervous System Balance, Health, and Well-being

2nd Edition

Stephen Elliott

With
Dee Edmonson, RN

The New Science of Breath
2nd Edition

Stephen Elliott

Copyright © 2005 Stephen Elliott
Copyright © 2006 Stephen Elliott

All rights reserved. No part of this publication may be reproduced, stored in a retrieval system or transmitted in any form or by any means, electronic, mechanical, audio, visual or otherwise, without prior written permission of the copyright owner. Nor can it be circulated in any form of binding or cover other than that in which it is published and without similar conditions including this condition being imposed on the subsequent purchaser.

ISBN 0-9786399-0-1

Cover design by
Stephen Elliott

Typesetting by
Robinson Associates

Published by
COHERENCE PRESS
1314 West McDermott Road
Suite 106, BPM 803
Allen, Texas 75013
email: inquiries@coherencepress.com

Reviews

"In the East it has long been known that breath is the cornerstone of health. *The New Science of Breath* offers the long awaited scientific explanation for this age old truth."
Dr. Meng-Sheng Lin, L.Ac.
VP Texas State Board of Acupuncture Examiners,
Honorary Surgeon General, The State of Texas

"I have just finished my second reading of your outstanding book *The New Science of Breath*. It was an educational experience for me and I came away from it with a resolve to incorporate a new method of breathing not only for my husband and myself, but for my patients as well. Overall, I am very impressed with the measure of optimism for taking conscious control of the autonomic nervous system via breathing."
Elsa Baehr, Ph.D., Clinical Associate, Department of Psychiatry
& Behavioral Sciences, Northwestern University Medical School,
Baehr & Baehr, LTD.

"In the rapid pace of modern civilization we have forgotten how to breathe. This *new science of breathing* shows us how "breathing coherently" can help restore a harmonious balance of mind, body, and spirit. Knowledge of breath, including an understanding of the physiology, the practice, and "the technology" is critical for anyone that teaches relaxation and stress management. As one who has employed Coherent Breathing in a clinical setting, I can attest to its effectiveness. In my view, this book should be required reading for all medical students."
Charles Lawe, Ph.D., Associate Director of Clinical Services,
Emory University Counseling Center

"As a rehabilitation psychologist and long-time biofeedback practitioner, I have found heart rate variability training to be a powerful addition to my practice and have been very impressed by the strong positive response of patients and clients to the "Breathing Pacemaker" compact disc RESPIRE 1. Elliott's newest contribution, *The New Science of Breath*, is likely to emerge as a standard reference for both the seasoned practitioner and clinicians who are new to the field. In one highly readable volume, he combines a scientifically rigorous yet readily understandable introduction to the complex topic of breath work and cardiovascular physiology, with chapters that are rich in practical training tips and strategies to jump-start the patient training experience. Among the half-dozen books on the topic I have read in recent years, it is among the most useful for the practicing clinician."

Roger Riss, Psy.D., Madonna Rehabilitation Hospital, Senior Fellow, Biofeedback Certification Institute of America

"I find your theories interesting. At times I am concerned about the softness of your evidence. On the other hand, postulations stimulate thought and controversy is necessary when putting forward new ideas."

Ronald DeMeersman, Ph.D., Columbia University Health Sciences

"Overall RESPIRE 1 seems like an excellent tool to assist someone with good steady breathing, facilitating a deep parasympathetic state. Personally, while breathing in synchrony with "Clock & Bell" I went into deep meditation."

Kenedy Singer, Ph.D. (Candidate), Neurofeedback Practitioner, 2005 Student Research Award Recipient, Annual Conference of the International Society for Neuronal Regulation (ISNR)

"I have been practicing yoga and martial arts for over 30 years and am always looking for ways to enhance my practice and improve my overall health and well-being. I have known Steve for over 10 years. He has been researching the subject of energy cultivation and optimal health for many years and has applied his deep knowledge of science and engineering to the discovery and development of the science of Coherent Breathing. I incorporated Coherent Breathing into my personal practice over a year ago and am experiencing major gains in depth and profundity."

Andrew Stronski – Martial Arts Colleague

This book is dedicated

to Yogi Ramacharaka (a.k.a. William Walker Atkinson), whose original work "Science of Breath", published 1905, has served as my impetus and inspiration.

I also dedicate this work to my editor, the late Adam Thomas (a.k.a. Den Robinson), protector of the English language.

It is also dedicated to those who have paved this way of knowledge.

Stephen Elliott, December 2005

Acknowledgements

I wish to thank Dee Edmonson for her invaluable contribution to this work. Her passion, enthusiasm, and personal commitment have elevated it to heights of which I could only dream.

I would also like to thank all of those who have taken it upon themselves to understand and put Coherent Breathing to the test, in particular Maria, Joseph, Steve, Dodie, Ryan, Janet, Jake, Erin, Ashley, Samantha, Ann, Andrew, Peter, and Jacob. I offer you my sincere thanks. May the practice of Coherent Breathing continue to serve you in every endeavor!

I would also like to express my gratitude to the professionals who afforded me their time, thoughtful consideration, and invaluable perspective by reading this work in advance, including Dr. Solyman Ashrafi, Dr. Elsa Baehr, Dr. Ronald DeMeersman, Dr. Daniel Galper, Dr. Charles Lawe, Dr. Meng-Sheng Lin, Dr. Roger Riss, and Dr. Billie Sahley.

I am also grateful to friends and relatives for their heartfelt enthusiasm, perspective, and criticism, often the kind that only real friends can offer, including Tony Alvares, Scott Bobo, Alan Boyce, Carol Elliott, George Elliott, Te Lin Lee, Michael McCarthy, Johnny Palka, Jim Schaefer, Andrew Stronski, and Ginger Vanhoose. And finally, to my immediate family, Kathleen, Evan, and Taylor, thank you for enduring the many long days, weeks, months, and years it has taken me to complete this work.

Special thanks to Larry Lewis whose deep scientific knowledge I have tapped many times during the course of developing this theory and its understanding. I also wish to thank Harry Rosenberg and Theo Klewansky of Agilent Technologies. Your excellent customer and application support made certain aspects of this research possible.

And to Michael Moore, our dialog has influenced this work in many ways.

Certain illustrations are by Georgia Minnich.

Sadgurunath Maharaj Ki Jay!

Author's Note

To the author's knowledge, the information set forth in this book is scientifically accurate and verifiable. It is based on the author's personal research and experience, as well as on insights gained from other relevant and/or expert sources. This being said, many of the thoughts and theories being put forward in this work have yet to be fully validated by others including recognized experts in the field.

The information offered herein is not intended to replace the services of a medical professional, nor to diagnose, treat, cure, or prevent any disease.

Coherent Breathing® is a physical breathing technique for the purpose of health enhancement. Those that choose to employ the method described herein should first consult a physician. The activities, physical and otherwise, while not known to pose any danger may, depending on individual health status, present a risk to some people. As such, neither the author of this book and of all accompanying products and documentation, nor his agents, nor his publishers, are responsible in any manner whatsoever for any injury that may occur as a consequence of use or misuse of information contained herein or in accompanying products and/or documentation. The sample group on which observations in this book are based includes 22 direct participants and approximately 125 indirect participants.

About the 2nd Edition

This second edition incorporates many small changes and one significant update/correction. The correction has to do with the physiological relationship between the heart rate variability (HRV) cycle and instantaneous arterial pressure. The first edition describes the arterial pressure wave during resonance, as varying with the heart rate variability cycle, peaks and valleys of the HRV cycle correlating directly with peaks and valleys of the arterial pressure wave.

While the aforementioned description is understood to be representative of the relationship that exists between heart rate and blood pressure during typical non-resonance, at resonance, heart rate variability and blood pressure cycles are 180 degrees out of phase, blood pressure falling as heart rate rises and visa versa. This revised understanding is based on the work of leading researchers, Vaschillo, et al., as reported in *Applied Psychophysiology and Biofeedback*, Volume 27, March 2002,[Note] and again referenced in Volume 31, June 2006.

I would like to thank Michael Duquette, CTRS, BCIAC, of the National Institutes of Health, Rehabilitation Medicine Department, for bringing the matter squarely to my attention. It comes at a time when, with the assistance of Carol Elliott, L.Ac., I have been studying "blood stagnation", a common symptomology of Traditional Chinese Medicine, attempting to understand its relationship to breathing. Reconciling the phase relationships between the heart rate variability cycle and arterial pressure has led to an important advancement in my own understanding that I have incorporated into the writing of the 2nd edition.

I would also like to thank the many readers of the first edition of *The New Science of Breath* for your strong interest, enthusiasm – and criticism. As of this writing, there are in excess of 1000 copies in circulation in half a dozen countries.

Note: Vaschillo, E., Lehrer, P., Rishe, N., Konstantinov, M., Heart Rate Variability Biofeedback As A Method For Assessing Baroreflex Function: A Preliminary Study of Resonance In The Cardiovascular System, *Applied Psychophysiology and Biofeedback*, Vol. 27, No. 1, 1-27 (2002).

Preface

The New Science of Breath proposes a revolutionary theory of health based on autonomic nervous system balance – *via breathing*. Eastern cultures have held breathing in high esteem for thousands of years; references to its importance are to be found in some of civilization's earliest writings. Yet even today, while modern biological and medical sciences have made great strides in exploring and understanding the details of the human organism, including respiration at the cellular level, beyond gas exchange the larger biological function of *breathing* remains something of a mystery and generally taken for granted.

If gas exchange is the first premise of breathing, *autonomic nervous system governance* is the second. The autonomic nervous system automatically governs subconscious bodily functions. As such, its action affects all aspects of life including health, well-being, and performance – mental and physical. Fundamental *homeostasis* requires autonomic balance; that is, equal sympathetic and parasympathetic action. Breathing, while not the only factor, literally governs autonomic balance, breathing frequency governing sympathetic nervous system emphasis, and breathing depth governing parasympathetic emphasis. Maximal cardiopulmonary resonance with a high degree of coherence, as evidenced by the heart rate variability cycle, is indicative of fine autonomic nervous system balance and optimality of breathing frequency and depth.

The New Science of Breath introduces Coherent Breathing,® an exciting approach to health enhancement. It is based on the premise that while at rest or semi-activity the adult cardiopulmonary system, inclusive of autonomic nervous system aspects, resonates at a specific frequency, this frequency being essentially the same for all adults. When the breathing frequency is consciously aligned with this "reference rhythm" with appropriate depth, it results in optimal autonomic nervous system balance. Autonomic nervous system balance yields mental and physical comfort, a positive emotional outlook, enhanced health and well-being, and improved biometrics.

Contents

Reviews..	iii
Acknowledgements.................................	vi
Author's Note...	vii
About the 2nd Edition.............................	viii
Preface..	ix
Foreword...	xii
Introduction...	xix
List of Illustrations..................................	xxii

Chapters

1.	Breathing – A Revised Premise..........	1
2.	Heart Rate Variability........................	8
3.	Cardiopulmonary Resonance.............	20
4.	Coherence..	27
5.	Coherent Breathing............................	31
6.	The Biometrics of Balance................	44
7.	Yoga and Meditation.........................	77
8.	Coherent Breathing – The Practice....	92
9.	Questions & Answers........................	101

Appendix A...	108
Notes...	114
References...	116
Index...	122
About the Authors...................................	130

Foreword

by Dee Edmonson, RN, Fellow BCIAC - EEG

We all have moments of great inspiration, a dream of making a profound difference in our chosen field or to positively affect others. Far too often the moment and the days slip away without seeing those goals or dreams realized. With **The New Science of Breath,** Stephen Elliott, life scientist, engineer, inventor, and esoteric artist, captures that moment, the result being a profoundly new and astonishing understanding of breath.

Seldom do we find a work that establishes a fundamentally new plateau of understanding – a new paradigm. This is such a book. With a practical orientation, yet profound theoretical depth, it synthesizes numerous complex concepts into a surprisingly simple yet incredibly elegant new understanding of the biological function of breathing. The theory is expressed with exceptional clarity and is supported with numerous powerful illustrations. I am confident that you will find, as I have, that it is an excellent tool for personal, professional, and educational use.

This is a book about breath – it is also a book about health and well-being – and about cardiopulmonary function – and psychophysiology – and neurology – and yoga and meditation. How can it possibly be about all of these things, you might ask. And the answer is, because it is fundamentally about autonomic nervous system balance, the matter of balance being central to all. While the ideas expressed are diverse, they reflect a unity that is profoundly more beautiful than any one concept in isolation, this beauty being a reflection of the human organism itself.

Over the past eleven months, I've had the opportunity to apply this "new science" in a clinical setting with well over 100 individuals and have observed first hand the profound impact it has had on their lives. I present twelve of these observations in chapter 6 – The Biometrics of Balance. Considering myself an "integrative" neurotherapist, I work with clients on a partnership basis incorporating conventional and alternative methods to maximize the body/mind's natural healing potential. I did not start out in the field of neurotherapy but have arrived here, since it affords me the greatest opportunity to *holistically* affect my client's well-being. Methods I employ include numerous forms of biofeedback, nutritional and medication education, stress mitigation, lifestyle modification, and *breathing*.

Steve and I met 5 years ago. I had just completed my first workshop in heart rate variability (HRV). While our backgrounds were diverse, we had a mutual interest in biofeedback. Steve was particularly interested in meditation and had been practicing EEG assisted meditation for some time. I introduced him to the computer-based HRV training program, he was fascinated, and that day he purchased one for his personal use.

What I know now, but did not know then, is that Steve is a highly disciplined life scientist in his own right, as well as an accomplished student and teacher of Eastern yogic and martial arts. He has a very intense interest in the workings of the human mind and body, and possesses a unique "systems view" resulting from a synthesis of diverse fields of knowledge including physiology, engineering, esoteric arts, and alternative medicine.

Over the course of growing to know each other, we shared our backgrounds, perspectives, and experiences. Steve revealed that while he had practiced yoga and martial arts for ten years beforehand, in 1985 he experienced a defining moment – kundalini Shakti[Note] entered his life "on her terms", guiding if not necessitating his spiritual pursuit and quest for knowledge. Since then, he has studied, contemplated, and practiced many systems of yoga, martial arts, and meditation, often in excess of 40 hours per week, his objective being to "coax the genie back into the bottle" – a feat that he never accomplished. Instead, with the aid of traditional Chinese medicine, he learned to live in harmony with this "paranormal" power.

Note: "Kundalini Shakti" literally means "coiled divine power" residing at the base of the spine. During a spiritual awakening, kundalini rises through the central channel to the crown of the head, piercing the chakras. At the crown, the individual soul merges with the supreme self and one attains self-realization. Kundalini Shakti is considered the supreme feminine force, the goal of yoga being the uniting of kundalini Shakti with Shiva – the supreme masculine force.

While Steve continued to research and practice both Eastern and Western methods, he says "I kept coming back to breathing – while I had surveyed and practiced most techniques both modern and ancient, I knew there was something extremely subtle that I was missing. It wasn't until I apprehended the phenomenon of cardiopulmonary resonance and understood its relationship to autonomic nervous system balance that I knew I'd found what I was looking for."

Some time after our initial meeting, Steve called and asked if I would help him conduct some research involving EEG and breathing. This being a keen interest of my own, I agreed and scheduled a number of uninterrupted hours to assist him in his experimentation. He explained that he had been exploring the "awakened mind" brainwave state for some time and was growing confident that exacting breathing was of central importance in eliciting the state. He had an audio CD that he had produced to aid him in "breathing coherently".

We did several "blind" HRV/EEG trials, each with his back turned to the computer monitor so that he was unable to view visual feedback. I noted that his HRV amplitude and coherence remained consistently high throughout the sessions, significantly higher and more coherent than I'd seen in his patterns previously. That and his EEG clearly demonstrated increased alpha amplitude with eyes open and closed. Afterwards, we discussed the results and I provided him with a copy of the data for his analysis. Thereafter, on occasion we would meet to "talk shop", my own interest in breathing and heart rate variability continuing to grow.

My own career began in 1968 in the field of cardiac care, both infant and adult, at New York University Medical Center where I worked in the cardiac medical/surgical intensive care unit. At that time, cardiac surgery was rapidly evolving and it offered a great opportunity for learning – the experience was both rewarding and challenging. However, my passion was in understanding the underlying dynamics of cardiac disorders leading to surgical intervention.

The cardiac catheterization laboratory, also at NYU, was expanding and needed a nurse with extensive cardiac experience. The physician in charge approached me and offered me the position of supervisor of the lab, which involved not only working with patients but also developing laboratory protocols, teaching personnel, managing equipment, and interacting with research and development teams.

This time was to have a profound impact on my life and career – the "cath lab" was a virtual gold mine of knowledge and experience. I had the opportunity to meet and work with many extraordinary physicians, fellows, students, surgeons, and mentors. At times I was student and at times teacher. While others found "crunching data", for example, reviewing hours of cineangiocardiographies for specific pathologies, to be boring and tedious, I appreciated every opportunity to increase my understanding of cardiac dynamics, intra-cardiac structure, and how abnormalities presented themselves in outward signs and symptoms.

Working with patients during catheterization, there were numerous opportunities to monitor their cardiopulmonary status as well as observe their psycho-physiological state. Patients and their families were often fearful of what at this time was a relatively new procedure, the patient's fear frequently manifesting itself in anxiety both before and during catheterization. It was not uncommon for patients exhibiting significant anxiety to experience everything from cardiac arrhythmia to spasm of the blood vessel at the site of the catheter insertion, resulting in delays or complications.

Noting the frequency of occurrence, I proposed that we develop an educational program to help patients and family members mitigate potential anxiety and be more comfortable with the procedure. One aspect of the program was breathing – using methods similar to Lamaze, I developed a technique that I referred to as "focus breathing". It involved selecting a spot on the wall or ceiling and having the patient focus their attention on that spot while breathing slowly and quietly. It usually helped – sometimes dramatically. Patients became more relaxed and less fearful and with it we had fewer complications. I could clearly see that breathing had a profound impact on patients both physically and psychologically... but what were the underlying mechanics?

I left NYU Medical Center to accept a unique position in a small hospital in eastern New Orleans, there to build a "university quality" catheterization laboratory "from the ground up". In addition to establishing the facility, the physicians wanted a program that would include staff and patient education. This provided me with a broader opportunity to bring to bear the special knowledge and skills that I had developed while at NYU, but most rewardingly, it offered an increased ability to positively influence patient outcome, both pre- and post- catheterization. Focus breathing continued to be a central aspect of patient education. And with it, my comprehension of the strong linkage between psycho-physiological state and breath continued to grow.

The logical evolution of my career led me into rehabilitation and ultimately to prevention of cardiac disease. Shortly after my relocation to Dallas in 1980, I served as a board member and speaker/educator for the Southwest Dallas County chapter of the American Heart Association. With my background in cardiac nursing, I was asked to focus on educating the professional community in the prevention of heart disease including diet, exercise, smoking cessation, and stress management. I emphasized a fifth tenet of well-being – use of breath to induce peace of mind and mitigate stress.

Two years later I was introduced to the work of Dr. E. Roy John, who was pioneering the Brain State Analyzer (BSA) and brain mapping techniques, and accepted his invitation to visit the research facility, again at NYU. I rapidly developed an interest in the technology and was fascinated by the concept of being able to explore psycho-physiological abnormalities non-invasively. With the encouragement of my mentors, I committed the following years to developing a strong working understanding of neurology and psychiatry in both clinical and home health settings.

In 1993, I began my internship in neurotherapy with a special interest in head injury and attention disorders, and went on to build my career in this field. I continued to emphasize to my clients the importance of breathing, but remained perplexed. Was there a *best way* to breathe – a way that would maximize health and well-being? If so, why was it not recognized or understood? I was once again reminded of the nagging question, how and why does breathing have such a powerful psycho-physiological effect in the first place? Could the effect simply be due to optimal perfusion?

About a year after our initial research sessions, Steve and I met again. By now his "new science" was starting to take shape. He was very confident that there was something extremely fundamental about the moment of cardiopulmonary resonance. He went on to explain that, not only had he been using "coherent breathing" during meditation, he had also incorporated it into his daily life as his normal breathing pattern, and that in both cases it had had a profound impact. I remember his words, "….when combined with stillness and relaxation, coherent breathing *is* meditation" and, "as I've employed it during my daily life it's yielded all the advantages one comes to expect of yoga, but on a continuous basis!"

Steve went on to explain that he was growing certain that, while not widely understood, *"cardiopulmonary resonance is a function of autonomic nervous system balance"*, and that *"autonomic balance is governed by breathing"*. The positive effects during both meditation and daily life "were simply the effects of residing in the state of ideal balance!" He related that he had validated his hypothesis using multiple biometric techniques and was convinced of its soundness. Needless to say, I was very intrigued – yet remained somewhat tentative. Was it possible that the *simple* act of breathing could govern autonomic nervous system status? Since our last meeting, my own understanding of the heart rate variability phenomenon had increased dramatically and based on my experience the theory made sense – my intuition was saying, "Yes!"

Steve explained that he was developing a series of "breathing pacemaker" products and would soon have them in prototype form. I related that I was very interested and volunteered to validate the therapeutic efficacy in a clinical setting. In early 2005, he provided me with a number of CDs and I began introducing "Coherent Breathing" to selected clients, typically as an adjunct to heart rate variability training. The initial results were profoundly positive – clients were able to achieve breakthroughs, both objectively and subjectively, often in a single session!

As the year progressed, I went on to employ Coherent Breathing with an increasing number of clients and conditions, typically in combination with other biofeedback methods including EEG, electrodermal response (EDR), and thermal response, and again the results continued to be extremely positive. By mid year I was completely convinced – breathing *is* a fundamental regulatory mechanism!

Steve and I continued to collaborate through his launch of Coherence and the writing of **The New Science of Breath**. When he invited me to author chapter 6, The Biometrics of Balance, I was delighted. This is a story that needs to be told. In telling it, I am completing my own circle of understanding, one that began decades ago. And with its completion, just like the cycle of breathing, a new circle of understanding is beginning…

Dee Edmonson

Introduction

We are at the brink of a revolution in our understanding of breathing and its relationship to human health and well-being. This "new science" quite possibly represents the most significant breakthrough in this understanding in recent times, for it promises a unifying theory of health based on autonomic nervous system balance, autonomic function underlying virtually every aspect of health and well-being.

The New Science of Breath contains information about breathing that is fundamentally new and exciting. The clear understanding on which the book is based has developed in just the past few years. Breathing has been revered as a means of health, well-being, and spiritual attainment, since time immemorial. This book will tell you why. It is also a how-to book as it will tell you how to achieve the advantages of optimal breathing for yourself and for your loved ones. You are encouraged to purchase Breathing Pacemaker® therapeutic audio CD, "RESPIRE 1" (visit www.coherence.com) so that you can begin to employ this knowledge as you read.

Breathing is a fascinating biological function that we all have in common. At the same time, it is a biological function about which most people know very little, for unless you are trying to accomplish something special, not much knowledge is required. "Breathing" pretty much takes care of itself.

Or so it seems……..

Not surprisingly, much of what we know about breathing has been known for thousands of years, this knowledge having been formulated long before oxygen, the life-sustaining quality of air, was identified circa 1800. Prior to the discovery of oxygen and the consequent understanding of its role in vertebrate physiology, there was irrefutable evidence that "life requires breath", for without breath, life, as the flame of the candle, will soon perish.

Previous popular books on the topic of breathing for health and well-being have tended to focus on the matter of optimizing the basic physiological function of gas exchange. Today, as it was a thousand years ago, the conventional wisdom is that it is preferable to breathe more slowly, more deeply, and more rhythmically. It is also generally accepted that it is preferable to breathe with more emphasis on the abdomen than on the chest. And lastly, if you have tried breathing while incorporating these factors, you will immediately find that it is not possible to do so without a final and necessary ingredient, this being "mindful" coordination. Each of these is an important tenet of correct breathing. But, as you are about to learn, this conventional understanding is far from complete.

If gas exchange is the first premise of breathing, *autonomic nervous system governance* is the second. The autonomic nervous system automatically governs subconscious (and most of them are subconscious) bodily functions.

As such, its action affects all aspects of life including health, well-being, and performance – mental and physical. Optimal homeostasis requires autonomic balance; that is, equal sympathetic and parasympathetic action. As you will see, autonomic balance is governed by the frequency and depth of breathing. *Optimal homeostasis is literally dependent upon correct breathing.*

As it turns out, while slower deeper breathing is generally preferable to more rapid shallow breathing, there is a frequency at which the human cardiopulmonary system, inclusive of autonomic nervous system aspects, naturally resonates. In **The New Science of Breath**, this frequency is referred to as "*The Fundamental Quiescent Rhythm*". This is the frequency at which the breathing rhythm equals the heart rate variability rhythm. (As you will learn in this book, the heart rate variability rhythm is the rate at which the heartbeat rate changes.)

Cardiopulmonary resonance is an indication of fine autonomic nervous system balance, the state wherein sympathetic and parasympathetic functions are of equal emphasis. It is also the state in which cardiac and pulmonary functions are operating at peak effectiveness and efficiency, the result being a *respiratory arterial pressure wave* of maximal amplitude that that rises and falls coincident with exhalation and inhalation, respectively – this wave being 180 degrees out of phase with that of the heart rhythm. A fact that has startling implications…..

The autonomic nervous system governs *automatic* bodily functions, those psychological and physiological functions over which we tend to have little conscious control, and to a large extent even those that we do. It is estimated that only .0001% of human function is actually "conscious", i.e. within our waking awareness, the remaining 99.9999% being subconscious! Consequently, the autonomic nervous system has a big job to do. This includes actions as subtle as the regulation of hormones, to actions as overt as the processing and elimination of waste. As such, the state of the autonomic nervous system underlies all psychological and physiological functioning, whether we are conscious of it or not. However, there is a bridge between our conscious mind and the subconscious action of the autonomic nervous system – breathing. Via breathing, we are able to effectively communicate our conscious desire for coherence, harmony, and wholeness.

With these comments, please pause, relax, read, muse, and – above all – breathe.

> *"What the bodily form depends on is breath (chi) and what breath relies upon is form. When the breath is perfect, the form is perfect (too). If breath is exhausted, then form dies. Therefore, the scholar who nourishes (his) life refines the form and nourishes (his) breath so as to nurture his life. No one has form without breath. Consequently, breath and form must be accomplished together. Isn't this evident?"*
>
> **The Master Great Nothing of Sung-Shan (circa 700 A.D.)**[1]

Note: "Chi" and "qi" refer to "the vital essence" in air as well as in the body, the latter being derived principally from breathing.

List of Illustrations

1. Autonomic nervous system
2. Bridges and related muscle groups
3. Dual control of diaphragm, intercostals and breathing bridge
4. Galvanic skin response measured in the hands as a function of breathing frequency
5. Sympathetic nervous system response in the hands when opening the eyelids wide and then relaxing them
6. HRV Amplitude ~ 14 beats. Average heartbeat rate = 78 beats per minute (BPM), moderate coherence
7. HRV Amplitude ~ 7 beats. Average heartbeat rate = 81 beats per minute, low coherence
8. HRV while breathing at 5 cycles per minute
9. HRV while breathing at 7.5 breaths per minute
10. HRV while breathing at 15 breaths per minute
11. HRV while breathing at 30 breaths per minute
12. Sinusoidal model comparing HRV amplitudes at 5, 7.5, 15, and 30 breaths per minute
13. A single heart rate variability cycle
14. A simplified Van Der Pol oscillator which approximates to that of vertebrate physiology
15. Heart rate variability spectrum while breathing at 5 cycles per minute
16. Spectral analysis of the relatively chaotic HRV of Figure 5
17. Spectral analysis of the low HRV of Figure 6
18. HRV while breathing at 4 breaths per minute
19. Some of the many varied activities of the autonomic nervous system coincident with the heart rate variability cycle

20. The mobius
21. Psycho-physiological correlates of autonomic nervous system balance
22. A pure sinewave
23. Fundamental quiescent rhythm and sympathetic/parasympathetic correlates
24. Inhalation/exhalation and changes at peaks and valleys
25. Shallow breathing immediately followed by deeper breathing (both at 5 cycles per minute)
26. 4-quadrant depiction of the relationship between frequency, depth and autonomic emphasis
27. Respiratory Arterial Pressure Wave and HRV cycle at cardiopulmonary resonance
28. Increase in biocurrent as a result of Coherent Breathing, relaxation, and stillness
29. The rise and fall of human "capacity" vs. age
30. Galvanic skin response measured in the left hand while clenching and then relaxing the right hand
31. The "Taiji Diagram" – representation of the Taiji paradigm
32. Examples of mantras and how they fit within the 6-second interval
33. Yoni Mudra
34. Bio-current of spontaneous "kriya" preceded by period of meditation employing Coherent Breathing
35. Sinewave model for Coherent Breathing
36. Pulse amplitude and blood volume measured during cardio-pulmonary resonance
37. A theory of cardiopulmonary operation at resonance
38. Thoracic cavity - source and sink
39. Mechanics of cardiopulmonary resonance - schematic view

1

Breathing – A Revised Premise

When we think of "breathing", we immediately associate it with the "oxygen imperative". This is very natural, due to the critical dependency we all have on taking that next breath. But, for most of us, this is where our understanding of breathing and its role in maintaining the human organism ends, and for good reason. While biological and medical sciences have made great strides in exploring and understanding the details of the human organism, including respiration at the cellular level, the larger biological function of breathing remains something of a mystery and is generally taken for granted.

The human organism consists of several trillion cells, these cells cooperating to form an incredibly elegant and sophisticated living system. Depending on its functional differentiation, each of these cells has a responsibility. In order to fulfill their respective roles, all cells require a few basics including oxygen, nutrition, a proper internal environment, and information/communication. While oxygen supply (and removal of carbon dioxide) is vitally important, the *function* of breathing has a role that extends far beyond that of mere "gas exchange", directly impacting both *milieu interieur* and information flow. It accomplishes this via its effect on the autonomic nervous system, the "operations and management" aspect of the central nervous system.

The vertebrate central nervous system (CNS) consists principally of autonomic and somatic subsystems. The autonomic subsystem presides over *automatic* functions, those over which

The New Science of Breath

we have little conscious control. This includes governance of organs and glands, and of internal processes, a few of which are circulation, digestion, metabolism, endocrine function, and breathing, although breathing is a special case. The somatic subsystem facilitates *conscious* control of skeletal muscle, yielding the ability to control body movement, for example walking.

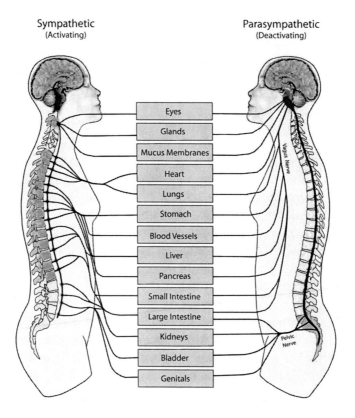

Figure 1: The autonomic nervous system

The autonomic subsystem generally consists of a bipolar arrangement wherein most organs receive a dual nerve supply, one pathway supplying sympathetic (activating) influence and the other supplying parasympathetic (deactivating) influence. In this way, the autonomic subsystem governs by both an "accelerator" and a "brake".

The New Science of Breath

The action of breathing is facilitated by the diaphragm and intercostal muscles. It is one of the few functions that is *explicitly* controlled by both autonomic and somatic subsystems, the autonomic function facilitating unconscious control, and the somatic function facilitating conscious control. The importance of this *dual control* is obvious. Without it, either we would have no ability to breathe consciously, for example hold the breath, a function that is vitally important to survival, or we would have to breathe consciously 100 percent of the time!

Chewing, blinking, swallowing, and excretory control are other functions for which the importance of dual autonomic and somatic control is self evident. Under normal circumstances (non fight, flight, freeze), conscious control takes precedence over unconscious control of these functions, and for this reason, we are able to chew, breathe, blink, etc., when and how we wish.

Generally, we have no direct conscious access to the autonomic nervous system, except via these points of *dual control* of which there are nine, breathing being the foremost. Each such point represents a *bridge* between somatic and autonomic subsystems that not only affords dual control, but facilitates conscious influence over the state of the autonomic nervous system at large! Note that the reason that *bridges* offer such an effective means of *communicating* with the autonomic nervous system is because, in order to function, we are <u>required</u> to exercise a fine degree of control over them! Bridges and related muscle groups are listed in Figure 2 from topmost to bottommost. Interestingly, they tend to possess an "open" and "closed" state.

Bridges: Related Muscles/Muscle Groups

1. Eyes: Orbicularis Oculi (closes), Levator Palpebræ Superioris (opens)
2. Jaw: Masseter (closes), Lateral Pterygoid (opens)
3. Tongue: Numerous muscles of tongue
4. Throat: Pharyngeal and Esophageal muscles (swallowing)
5. Hands: Numerous muscles of forearm and hand
6. ***Breathing: Diaphragm and Intercostals***
7. Urinary: Urethral Sphincter with numerous pelvic muscles
8. Excretory: Anal Sphincter with numerous pelvic muscles
9. Feet: Numerous muscles of lower leg and foot

Figure 2: Bridges and related muscle groups

Bridges may aid or detract from autonomic balance. By "tensing" a bridge we are able to facilitate a shift toward sympathetic emphasis. Conversely, by "relaxing" a bridge we are able to facilitate a shift toward parasympathetic emphasis. As it relates to the *breathing bridge*, increasing breathing frequency results in an increase in sympathetic emphasis – decreasing breathing frequency and increasing breathing depth facilitates an increase in parasympathetic emphasis.

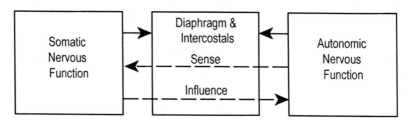

Figure 3: Dual control of diaphragm, intercostals, and the breathing bridge

Figure 3 is a logical representation of dual control of the diaphragm and intercostal muscle groups, and the ability for somatic control to both "sense" and "influence" the state of the autonomic nervous system. This effect can be verified via galvanic skin response (GSR), a useful indicator of relative sympathetic nervous system status. Figure 4 on the next page depicts GSR (body resistance, measured in ohms) while breathing at 5 breaths per minute, shifting abruptly to 15 breaths per minute, and then shifting back to 5 breaths per minute. Lower resistance is indicative of sympathetic emphasis and higher resistance is indicative of sympathetic withdrawal.

It can be seen that while breathing at the rate of 5 cycles per minute, resistance averages 60,000 ohms. Coincident with the shift in breathing frequency from 5 to 15 breaths per minute (the typical adult breathing frequency), resistance drops sharply to 46,000 ohms and remains there until the breathing frequency changes. At the end of 6 minutes, the breathing frequency once again abruptly changes to 5 cycles per minute, at which time resistance drops momentarily and then begins climbing. At the end of 9 minutes, the resistance value is 52,000 ohms and climbing steeply, indicating sympathetic withdrawal.

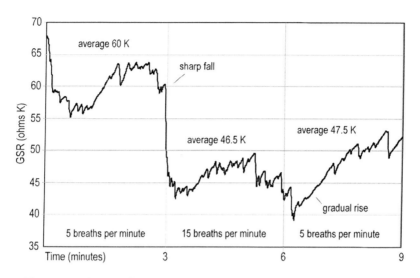

Figure 4: Galvanic skin response measured in the hands as a function of breathing frequency

Opening (wide) and then closing the eyes, clenching and relaxing the jaw, and flexing and relaxing the perineum (pelvic floor) yield the same effect, as does clenching and relaxing the hands and feet, although the hands and feet do not possess the same obvious degree of *automatic* autonomic nervous system governance. As with breathing, the effect of each of these bridges on the sympathetic nervous system can be verified biometrically. Figure 5, overleaf, demonstrates the rapid sympathetic nervous system response in the hands each time the eyes are opened wide. Here again, a drop in resistance indicates sympathetic emphasis and a rise in resistance indicates sympathetic withdrawal.

Dual control affords us the ability to consciously moderate the state of the autonomic nervous system at large, not just the function at hand; for example, holding the breath when we go under water or closing the eyes in a dust storm. Over and above this, it presents a means of consciously influencing or guiding relative sympathetic/parasympathetic emphasis.

But why should the autonomic nervous system require guidance? Does it not possess the innate capability to govern optimally on our behalf? The answer is "yes" and "no". The autonomic nervous system is expert at handling urgent matters, those that threaten our existence. It fully accepts the "survival imperative" as its *raison d'être*. So much so, that even during normal (non fight, flight, freeze) circumstances it tends toward survival readiness.

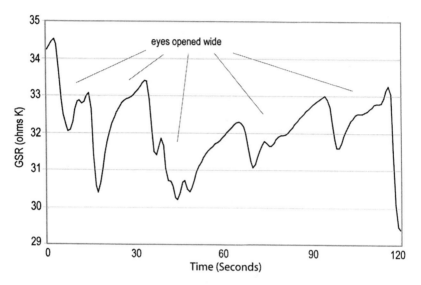

Figure 5: Sympathetic nervous system response in the hands when opening the eyelids wide and then relaxing them

Generally, survival involves sympathetic nervous system emphasis, including production of the biotransmitters cortisol and epinephrine, quickening of the heart rate, raising the blood pressure, boosting energy production, flexing of skeletal muscle, slowing of digestion, and many other subtle psychological and physiological adjustments. Unfortunately, even after the sympathetic stimulus is removed, the autonomic nervous system tends to remain on heightened alert. For this reason, most of us live with elevated heart rates, incessant skeletal muscle tension, and some degree of anxiety.

Evidence of sympathetic bias can be detected relative to these dual control points – they tend to remain somewhat "flexed". As it relates to breathing, this means that the frequency of breathing tends to remain relatively high and the depth relatively low as compared to that of ideal homeostasis –

that is, sympathetic/parasympathetic balance. Because we can readily sense the status of these points, they serve as both "thermometer" and "thermostat". This is to say that they are useful for both gauging and guiding autonomic nervous system behavior.

The emerging medical understanding is that many modern-day maladies are rooted in autonomic nervous system imbalance, specifically, "sympathetic dominance". Sympathetic dominance is the condition wherein the autonomic nervous system persists in the state of relative sympathetic emphasis and parasympathetic under-emphasis. The effect of this is that bodily systems, including psychological and physiological functions, remain in a constant state of hyperactivity. Common maladies resulting from sympathetic dominance include elevated heart rate, high blood pressure, headache, temporomandibular joint dysfunction (TMJ), general muscle tension, pain, anxiety, worry, digestive problems, incessant mental chatter, attention deficiency (inability to concentrate), a lack of energy, and sleeping problems. It is important to note that the state of the autonomic nervous system affects every cell in the body! If the autonomic nervous system is on heightened alert, every cell is on heightened alert.

Aging is primarily a function of bodily systems wearing out. Because sympathetic dominance results in persistent over-activation of all bodily systems it contributes substantially to the aging process. This is specifically true of, but not limited to, the cardiovascular system, where chronic excessive pressure over time literally destroys the heart, the blood vessels, and the kidneys.

Is this the inevitable plight of modern man? Are we *programmed* to suffer the ravages of sympathetic over-emphasis? For now, yes… unless we assume responsibility for our state of being. **The autonomic nervous system alone does not possess the ability to govern optimally without our conscious participation!** Humans are endowed with consciousness, and we can bring consciousness to bear on the matter of balance.

2

Heart Rate Variability

The human autonomic nervous system takes its cue from the frequency and depth of breathing. The typical adult breathes at the rate of 15 breaths per minute, a rate that the autonomic nervous system interprets as "fight or flight". In other words, even though you may be sitting quite still reading this page, if you are breathing at the typical rate of 15 breaths per minute, your autonomic nervous system is shifted toward "sympathetic" emphasis. Generally, sympathetic emphasis involves activation of bodily systems resulting in increased heartbeat rate, increased muscle tension, and the production of biochemicals associated with *threatened survival*. Because most adults breathe at this rate much of the time, most adults persist in a relative state of chronic "sympathetic dominance". As previously mentioned, the emerging medical understanding is that many modern-day maladies are rooted in autonomic imbalance. What has not been understood is the root cause of this imbalance. This *new science* asserts the theory that the primary cause of autonomic imbalance is in fact suboptimal breathing. The good news is that just as suboptimal breathing results in imbalance, optimal breathing results in balance. Consequently, via proper breathing, sympathetic dominance and its myriad affects may be averted.

To understand this more clearly, it is necessary to introduce "heart rate variability" or "HRV". Heart rate variability is the rate at which the heartbeat rate changes. It is very important to us because, at present, it is the only non-invasive window we have into the status and operation of the autonomic nervous system – parasympathetic function being particularly difficult to observe other than via HRV. Heart rate variability can be monitored very effectively, especially with present day technology.

Most of us are familiar with "heartbeat rate" as it is one of the most common biometrics. Your doctor probably assesses your heartbeat rate and blood pressure with each visit. The conventional view of a "normal" heartbeat rate is 72 beats per minute. If you examine the heartbeat more carefully, as can be done using an electrocardiograph, a heart rate variability monitor, or even a stethoscope, it can be seen that the "beat" varies in time. This variation is governed by a complex interplay between cardiopulmonary mechanics and the autonomic nervous system.

Heart rate variability is a highly intricate, non-linear, dynamical phenomenon possessing dozens of complex characteristics and requiring an equal number of sophisticated metrics to adequately describe it in all of its complexity. This having been said, for the purposes of the ensuing discussion we need only concern ourselves with its most apparent attributes, these being amplitude, frequency, average heartbeat rate, and overall "coherence".

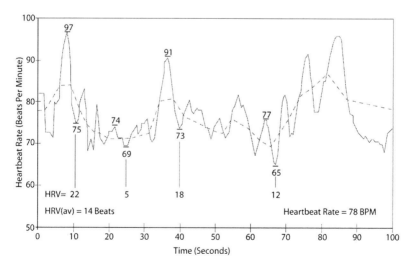

Figure 6: HRV amplitude ~ 14 beats. Average heartbeat rate = 78 beats per minute (BPM), discernible coherence

The New Science of Breath

Amplitude is indicative of the "tone" and "responsiveness" of the cardiopulmonary system including autonomic nervous system aspects; frequency is the rate at which the heartbeat rate changes; average heartbeat rate is indicative of relative sympathetic/parasympathetic status; and "coherence" is a measure of overall regularity and consistency of the heart rate variability cycle. Figure 6, on the previous page, presents an example of heart rate variability measured over a period of 100 seconds. HRV amplitude equals the highest heartbeat rate, represented by the peaks, minus the lowest heartbeat rate, represented by the valleys. To determine average HRV it is necessary to subtract the average of valley values from the average of peak values. For simplicity, we will average just a few of the points, those with numeric values. This yields an average peak of 85 and an average valley of 71. Subtracting the average valley from the average peak yields an $HRV_{(av)}$ of 14 beats. In other words, for the purpose of this exercise, the average difference between the highest heartbeat rate and the lowest heartbeat rate for this 100-second period is 14 beats. The dotted line represents the approximate mean as it changes in time.

Whether you know it or not, your doctor assesses your heart rate variability when he or she places the stethoscope on your chest and asks you to breathe deeply. Not only is your doctor listening to your heartbeat and the air as it passes in and out of your lungs, but is also listening to the variability of your heartbeat as you inhale and exhale. Is it increasing as you inhale? Is it decreasing as you exhale?

While science recognized the existence of HRV as early as the 1700s, until the advent of powerful computing the tools did not exist to adequately characterize its role and significance in vertebrate physiology. However, several decades of research have now been undertaken with the aid of contemporary computing, in particular that of computer aided real time fast Fourier transformation. At present, HRV analysis is an automated aspect of state-of-the-art cardiological assessment. Via many independent research studies, heart rate variability is becoming well established as a reliable indicator of stress level and age, HRV amplitude varying inversely with both factors. Low HRV amplitude correlates highly with elevated risk of cardiac sudden death, coronary heart disease, and mortality resulting from *all causes*. In other words, *heart rate variability is a key indicator of the overall integrity and well-being of the human organism.*

The physiological correlate of stress is "sympathetic emphasis". As the burden of stress or "allostatic load", the cost of excessive adaptation,[2] increases, the human organism responds by strengthening its *defenses*. Specifically, this involves ratcheting up sympathetic nervous system emphasis. The strain we feel when we are "stressed out" is the strain that results from the sustained increase in psychological and physiological action resulting from sustained sympathetic emphasis.

Breathing is intimately linked to heartbeat rate via a phenomenon known as "respiratory sinus arrhythmia" (RSA). This relationship is such that heartbeat rate tends to increase coincident with inhalation and decrease coincident with exhalation. As such, the amplitude and frequency of the heart rate variability pattern relate strongly to the frequency and depth of respiration. *Medical Physiology* describes the heartbeat rate as a function of respiratory sinus arrhythmia as varying no more than 5% during resting respiration and by up to 30% during deep respiration.[3] This author posits that 5% is in fact characteristic of typical adult breathing while in the state of rest, i.e. typical adult breathing is both shallow and relatively fast.

HRV signatures may range widely between people of the same age, some exhibiting low amplitude and low coherence while others exhibit high amplitude and high coherence. Figures 7 and 8 are representative of these extremes. Having said this, it is very unusual that one exhibits high amplitude and high coherence without prior knowledge or training.

HRV amplitude correlates highly with stress level cum sympathetic emphasis – the higher the sympathetic emphasis, the lower the heart rate variability amplitude. Why is this so? This *new science* asserts the answer: Sub-optimal breathing! The frequency and depth of breathing vary with stress, and heart rate variability amplitude varies with breathing. In other words, when under stress, people tend to breathe rapidly and shallowly. It is well established that acceleration of breathing is one of the body's first reactions to increased stress. Breathing accelerates to fuel the increased energy production required for survival,

The New Science of Breath

energy production necessitating the reaction of oxygen with nutritional constituents and the exhaust of increased levels of carbon dioxide. Stress causes rapid breathing – rapid breathing results in sympathetic emphasis, thus forming a vicious circle.

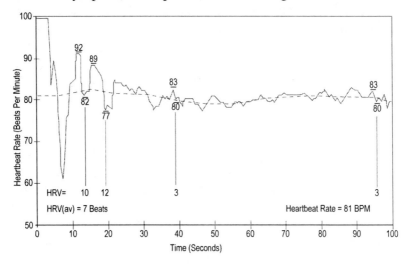

Figure 7: HRV amplitude ~ 7 beats. Average heartbeat rate = 81 beats per minute, low coherence

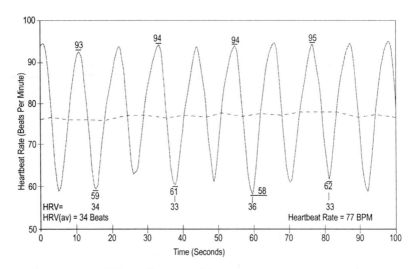

Figure 8: HRV while breathing at 5 cycles per minute amplitude ~ 34 beats. Average heartbeat rate = 77 beats per minute, high coherence

Figure 8, above, is a view of HRV while breathing at the rate of 5 cycles per minute. Figures 9, 10, and 11 are snapshots of the same individual consciously breathing at 3 additional frequencies, 7.5, 15, and 30 breaths per minute, respectively.

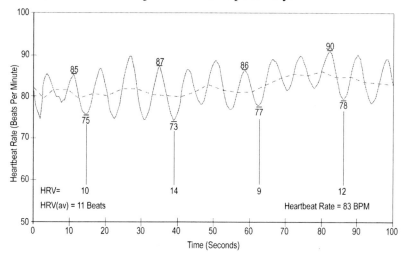

Figure 9: HRV while breathing at 7.5 breaths per minute, amplitude ~ 11 beats. Average heartbeat rate = 83 beats per minute, relatively high coherence

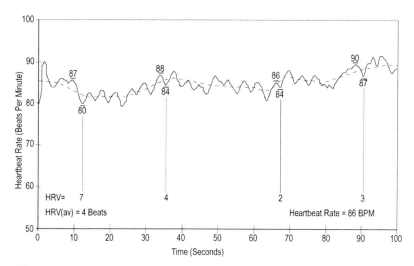

Figure 10: HRV while breathing at 15 breaths per minute, amplitude ~ 4 beats. Average heartbeat rate = 86 beats per minute, modest coherence

The New Science of Breath

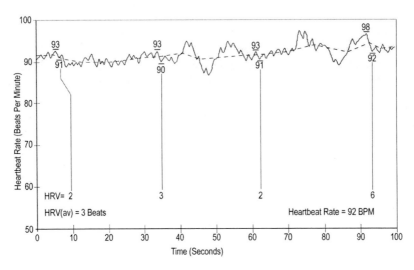

Figure 11: HRV while breathing at 30 breaths per minute, amplitude ~ 3 beats. Average heartbeat rate = 92 beats per minute, modest coherence

Figure 12, on the following page, presents a sinusoidal model which approximates to the HRVs of figures 8, 9, 10, and 11, these being represented by the lowermost to the uppermost curves, respectively. Again, these are HRV signatures of the same person deliberately breathing at 4 different frequencies and otherwise seated and at rest.

Relative to Figure 12, a few things are immediately apparent:

1) HRV amplitude varies radically with breathing frequency. For example, HRV amplitude at 5 breaths per minute is 34 beats. HRV amplitude at 7.5 breaths per minute is 11 beats. A change of 2.5 breaths per minute yields a change in amplitude of 23 beats!

2) This change in amplitude is a consequence of a rapidly diminishing HRV low or valley. In other words, the valley shifts upward dramatically with increased breathing frequency.

3) As a consequence, the average heartbeat rate also shifts upward as breathing frequency increases. Average heartbeat rate is indicative of relative sympathetic/parasympathetic emphasis.

4) HRV frequency varies with breathing frequency. This is true as long as the breathing frequency is synchronous and relatively coherent.

The New Science of Breath

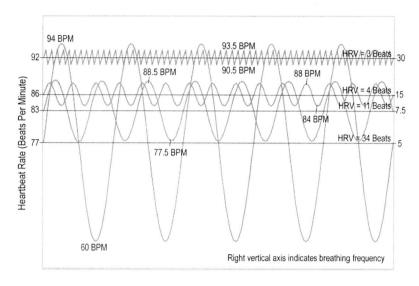

Figure 12: Sinusoidal model comparing HRV amplitudes at 5, 7.5, 15, and 30 breaths per minute

Let us step back from this for a moment and get our bearings. If we look at a single HRV cycle as is depicted in Figure 13 overleaf, the uppermost peak is indicative of real time sympathetic activity. The bottommost valley is indicative of real time parasympathetic activity. The average heartbeat, the mathematical mean, is indicative of relative sympathetic/parasympathetic bias. For a highly coherent HRV cycle, the average heartbeat rate relates to HRV amplitude in this way:

$$\text{Average heartbeat rate} = \text{HRV}_{(peak)} - [(\text{HRV}_{(peak)} - \text{HRV}_{(valley)})/2], \text{ or}$$
$$= \text{HRV}_{(valley)} + [(\text{HRV}_{(peak)} - \text{HRV}_{(valley)})/2]$$

A shift upward of the positive peak is indicative of increased sympathetic emphasis. A shift downward of the positive peak is indicative of decreased sympathetic emphasis. A shift downward of the valley is indicative of increased parasympathetic emphasis. An upward shift of the valley is indicative of decreased parasympathetic emphasis.

With this in mind, we can now see that breathing at 5 cycles per minute yields *both* the most sympathetically and parasympathetically inclined heart rate variability cycle.

The New Science of Breath

Breathing at 30 cycles per minute yields the most sympathetically and least parasympathetically inclined HRV cycle. Breathing frequencies in between tend toward increased sympathetic emphasis and decreased parasympathetic emphasis. The typical adult breathes at a rate commensurate with the 3rd curve from the bottom, 15 breaths per minute.

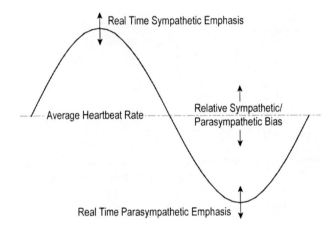

Figure 13: A single heart rate variability cycle

The critical issue is not that more rapid breathing results in increased sympathetic emphasis – which it does. The critical issue is that *rapid breathing strongly negates parasympathetic emphasis*. If we once again examine the HRV corresponding to 5 breathing cycles per minute, we see that its sympathetic aspect (peak) is just as pronounced as that of 30 breaths per minute, both of which approximate to 93-95 beats per minute. On the other hand, the parasympathetic aspect (valley), relating to 30 breaths per minute as compared to 5 breaths per minute diminishes by over 30 beats! *Here we can see one reason why heart rate variability amplitude is important to health and well being – it equates to autonomic balance!*

If we accept that the valley represents the moment of maximal cardiopulmonary rest and relaxation, then it is this rest period that is sacrificed. Given that this "signal" is distributed throughout the body via sympathetic and parasympathetic nervous structures, this rest period is sacrificed relative to the entire organism, i.e. muscles, organs, glands, etc.

It is very important to remember that the measurements behind the model of Figure 12 were taken while the subject was seated and at rest with the exception of consciously breathing at different rates. This observation makes clear the fact that the cardiopulmonary system, inclusive of central nervous system aspects, takes its cue from the breathing rate, i.e. the faster the breathing cycle, the faster the average heartbeat rate and the lower the heart rate variability amplitude. In this regard, the behavior of the autonomic nervous system while at rest is very similar to its behavior during exercise, i.e. breathing frequency represents an exercise-like stimulus.

Instantaneous arterial pressure follows respiration resulting in the phenomenon known as the *"respiratory arterial pressure wave"*. Calling on Medical Physiology, during rest, the 5% variation in heartbeat rate corresponds to an arterial pressure change of 4 to 6 mmHg (millimeters of mercury); during deep respiration the 30% variation in heartbeat rate corresponds to changes in arterial pressure of up to 20 mmHg.[4]

Initially, is seems as though the heart rate variability cycle is the key determinant of instantaneous pressure – and this is true under specific circumstances. Instantaneous pressure aligns closely in time with the heart rate variability cycle when it is low in amplitude, for example when there is a 5% variation in heartbeat rate. Alternatively, when HRV amplitude is high, instantaneous pressure follows respiration, arterial pressure and heart rate variability cycles varying by 180 degrees.[5] How and why does this phase difference exist between the HRV cycle and the arterial pressure wave as breathing becomes slower and deeper? (Please refer to Appendix A, Figure 37 for this discussion.)

Inhalation results in a strong negative thoracic pressure (a vacuum) resulting in expansion of the lungs and the inflow of air from the external environment. Simultaneously, this internal pressure differential causes blood vessels in the chest to expand. Because the chest is replete with blood vessels, this results in significant "storage" of blood in the chest during inhalation, thereby reducing blood flow returning to the heart, lowering heart output and arterial pressure.[6] The opposite occurs during exhalation. In this way, breathing itself acts to raise and lower blood pressure, resulting in the *respiratory arterial pressure wave* which rises and falls with exhalation and inhalation, respectively.

The respiratory arterial pressure wave can be observed using a plethysmograph, where we can see pulse amplitude and blood volume in the extremities rising and falling with exhalation and inhalation, respectively. (See Figure 36) Because respiratory sinus arrhythmia has an inverse relationship with breathing to that of arterial pressure, we see that heart rate variability and the arterial pressure wave are 180 degrees out of phase, i.e. upon inhalation heart rate increases yet arterial pressure falls and visa versa.

From this we see that breathing plays a *key* part in moving the blood. It appears that when breathing is relatively fast, shallow, and asynchronous (Figure 7), the heart effectively bears the burden of generating and sustaining blood pressure, and for this reason, arterial pressure aligns with the HRV cycle. As breathing becomes slower and deeper (Figure 8), the mechanical action of breathing itself takes on a greater and greater role in moving the blood. And, as a consequence, heart rate increases dramatically during inhalation and decreases dramatically during exhalation. Note that if heart rate did not increase and decrease accordingly, arterial pressure would fall and rise excessively. In this way, the HRV cycle plays a key role in moderating the *respiratory arterial pressure wave*. Building on this understanding, what would the consequence be of very low HRV amplitude, i.e. the heart bearing the burden of generating and sustaining blood pressure?

It is generally accepted that sympathetic emphasis results in simultaneous increased heartbeat rate, heart output, and cardiovascular constriction leading to elevated arterial pressure. This is the means by which the autonomic nervous system manages blood pressure, although potentially *pathologically*, in the absence of breathing depth and the resulting respiratory arterial pressure wave. It is also implicated as the primary physiological mechanism behind hypertension, hypertension being a sympathetically mediated disease. (The common approach to managing hypertension is via drugs that moderate sympathetic bias, reduce heart output, and reduce arterial constriction.) Given that the typical adult (at sea level) breathes between 10 and 20 times per minute, does suboptimal breathing not play a significant part in the present hypertension pandemic? Of course it does!

As previously discussed, the autonomic nervous system has a tendency to err on the side of sympathetic emphasis, not returning on its own to the state of sympathetic/parasympathetic balance without our conscious involvement. Oddly enough, while it seems implausible that we would get in an automobile and expect to drive it without conscious involvement, this is the nature of the relationship many of us tend to have with our physical bodies. Just as we experience varying road and weather conditions while driving, our bodies/minds are confronted with varying situations that require no less conscious engagement. It makes no more sense for your autonomic nervous system to be racing when you are sitting behind the steering wheel at a red light than it does for the engine of your car. In both cases, inappropriate acceleration wastes energy and creates unnecessary wear and tear on the vehicle.

However, we tend not to be entirely aware that "racing" is going on. We feel it because it creates mental and physical discomfort, but often we don't know why we are feeling discomfort. Frequently we attribute it to some other "cause", ultimately increasing stress and discomfort.

Building on the analogy of the automobile, it would be nice if our autonomic nervous system functioned like an automatic vs. a stick shift and by and large it does. It goes from 0-120 M.P.H. very nicely on its own. However, it requires us to downshift if we want to slow down to an "idle".

Downshifting *requires our conscious involvement.* Specifically it requires us to use our somatic nervous system to influence our autonomic nervous system to slow down. The primary way that this is accomplished is via the *breathing bridge*, i.e. by consciously slowing breathing frequency and increasing breathing depth. Consciously relaxing other *bridges* aids in the process. When you practice this consistently over a period of days and weeks, you are training your mind to perform these actions for you. This is like driving once you've had enough experience that it becomes automatic. At first it takes cautious deliberation. But once you've trained yourself, all it takes is "an ounce of mindfulness".

3

Cardiopulmonary Resonance

The vertebrate cardiopulmonary system inclusive of central nervous system aspects contains an "oscillator". R.I. Kitney of Imperial College, London, asserted this general concept in his paper to the First International Symposium on Cardiovascular Respiratory and Somatic Integration in Psychophysiology, held in 1983.[7] This oscillating function is usually described in terms of a "Van Der Pol oscillator", named after its inventor, Balthazar Van Der Pol (1889-1959), a Dutch electrical engineer and pioneer in nonlinear dynamics as well as dynamics of the heart.

We don't need to know too much about the details of this oscillator except that it aspires to resonate at its natural center frequency and that via its "sync" input, it will synchronize with and ultimately phase-lock to the respiratory signal if that signal is both synchronous and of sufficient magnitude. In this way, the respiratory cycle *may* either push or pull the oscillator's output frequency and phase.[8]

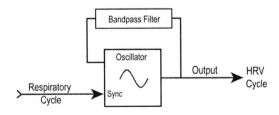

Figure 14: A simplified Van Der Pol oscillator which approximates to that of vertebrate physiology

The output of this theoretical oscillator is the signal we know as heart rate variability (HRV). If we sweep the sync input with varying breathing frequencies, as was done to yield the graph of Figure 12, and look for the moment of maximal output amplitude, we can determine the natural center frequency of the oscillator. As it turns out, this natural center frequency while the body is erect and in a state of rest or semi-activity, for example sitting at a desk working, is .085 hertz and is essentially the same for all adults. Therefore, breathing at .085 hertz with commensurate depth yields an HRV signature with maximal amplitude. In *The New Science of Breath*, this frequency is referred to as *The Fundamental Quiescent Rhythm*, the rhythm at which the cardiopulmonary system, inclusive of autonomic nervous system aspects, naturally resonates.

Many will note that the resonant center frequency of .085 hertz is different from the generally accepted figure of .1 hertz. Whereas .1 hertz translates to 6 breaths per minute, .085 hertz translates to ~5 breaths per minute. The reason for this difference is that .085 hertz assumes respiratory sinus arrhythmia (RSA) "tone", this tone being developed via significant practice and incorporation of slow deep breathing into one's daily life. This, and learning to "sense" and "relax" the *bridges* including the diaphragm and intercostals. Again, the reason for this is that, if tense, bridges will inhibit parasympathetic prominence, preventing autonomic balance.

As mentioned previously, HRV is a window into autonomic nervous system functioning and the only practical non-invasive means to observe this intrinsic autonomic nervous system rhythm. Figure 15 presents a spectral analysis of the heart rate variability signal while breathing at the *Fundamental Quiescent Rhythm*, plotting HRV frequency vs. power as averaged over a period of interest. This spectral signature is highly symmetrical, demonstrating that the associated HRV signal possesses a high degree of spectral purity, frequencies being largely contained to a relatively narrow band around the center frequency of .085 cycles per second. This narrow band and high symmetry demonstrates that the quality factor or "Q" of the theoretical oscillator is quite high, lending further credence to its existence in fact.

The New Science of Breath

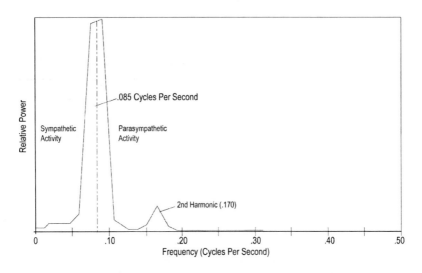

Figure 15: Heart rate variability spectrum while breathing at 5 cycles per minute

This spectral signature is of special significance because it is representative of optimal autonomic nervous system balance, spectral components to the left of .085 hertz being primarily reflective of sympathetic activity, and spectral components to the right of .085 hertz being primarily reflective of parasympathetic activity. In this case, with the exception of equal frequency components to the immediate left and right of the "monument" centerline, there are neither. (The secondary peak representing the second harmonic, .17 hertz, is a vestige resulting from the digital signal processing-based measurement method.) This spectral signature approximates to that of a pure sine wave, which due to its spectral purity would appear more as a "needle" than a "monument".

Contrast Figure 15 with Figures 16 and 17, the spectral analyses of the chaotic HRV cycle of Figure 6 and the low HRV cycle of Figure 7.

Both demonstrate strong sympathetic emphasis, frequency components to the left of the dotted line being primarily indicative of sympathetic nervous system action and those to the right being primarily indicative of parasympathetic action. The HRV of Figure 16 demonstrates a higher degree of parasympathetic activity than that of Figure 17. The HRV of

Figure 17 demonstrates a higher degree of sympathetic action than that of Figure 16.

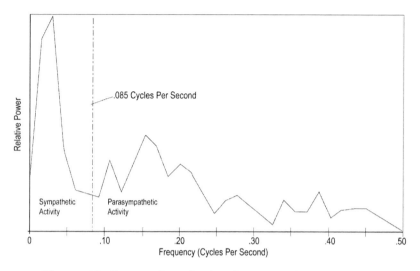

Figure 16: Spectral analysis of the relatively chaotic HRV of Figure 6

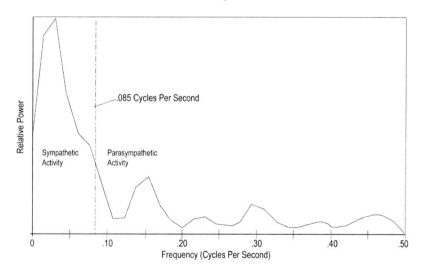

Figure 17: Spectral analysis of the low HRV of Figure 7

The frequency .085 hertz yields a period of 11.76 seconds for a breathing rate of 5 cycles in ~60 seconds (58.8 to be exact). If

The New Science of Breath

we increase the breathing frequency above 5 cycles per minute, HRV amplitude will fall rapidly. If we decrease the breathing frequency below 5 cycles per minute, the HRV waveform will diminish in amplitude and will become increasingly distorted and incoherent. Relative to the oscillator model, this is due to an inability of the oscillator to track and phase-lock with breathing frequencies below that of resonance (given the specific circumstance).

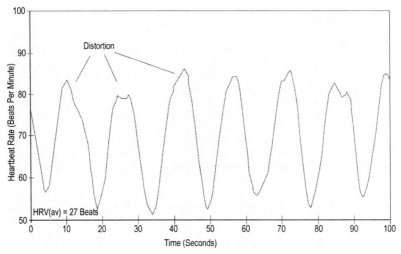

Figure 18: HRV while breathing at 4 breaths per minute

The autonomic nervous system coordinates numerous activities in synchrony with the heart rate variability cycle. Figure 19 overleaf, depicts these activities (at resonance) in the context of a pendulum, the pendulum being a simple sinusoidal oscillator.

Upon inhalation, thoracic pressure becomes negative (a vacuum). The magnitude of this pressure depends on the extent of inhalation. Blood vessels in the chest expand storing increased blood. Bronchi dilate and oxygen is transferred into the bloodstream. Due to storage of blood in the chest, arterial pressure as measured at the extremities falls. However, heart rate increases consistent with respiratory sinus arrhythmia. Arterial walls constrict as heart rate increases, thereby limiting the decrease in peripheral arterial pressure.

Upon exhalation, thoracic pressure becomes positive, the magnitude of this positive pressure again depending on the extent of exhalation. Due to increased pressure, blood vessels in the chest shrink increasing blood flow to the heart. However, heart rate decreases consistent with RSA. Arterial walls relax as heart rate decreases, thereby limiting the increase in peripheral arterial pressure. While venous blood flow occurs on a continuous basis, venous reservoirs are filled coincident with increasing arterial pressure and empty coincident with decreasing arterial pressure.

Sympathetic Phase
inhalation
negative thoracic pressure
bronchi dilate
exchange of O_2 and CO_2 maximized
blood vessels in the chest expand
arterial pressure decreases
heartbeat rate increases
arterial constriction occurs
net arterial blood flow decreases
peripheral arterial pressure decreases
venous blood flow accelerates
venous reservoirs empty

Parasympathetic Phase
exhalation
positive thoracic pressure
bronchi constrict
exchange of O_2 and CO_2 minimized
blood vessels in the chest shrink
arterial pressure increases
heartbeat rate decreases
arterial relaxation occurs
net arterial blood flow increases
peripheral arterial pressure increases
arterial blood flow accelerates
venous reservoirs fill

Figure 19: Some of the many varied activities of the autonomic nervous system coincident with the heart rate variability cycle (depicted at resonance)

Upon consideration, it may be recognized that this relationship is ideal relative to the cardiopulmonary imperative, this being the alternating supply of oxygen and removal of carbon dioxide from body tissues, as well as the delicate maintenance of arterial pressure and blood pH.

The New Science of Breath

You will recall that during times of deep respiration, the arterial pressure wave can vary by as much as 20 millimeters of mercury (mmHg). During cardiopulmonary resonance, I have measured arterial pressure waves in excess of 60mmHg. That is to say that the difference in arterial pressure between the peak and valley of the wave is greater than 60mmHg. This is somewhat astounding and makes clear that conventional blood pressure measurement is only providing a small part of the total arterial pressure picture. In fact, it may be argued that conventional blood pressure measurement assumes that the respiratory arterial pressure wave is either non-existent or very small, and for the general populace this would be fitting. It also demonstrates that *during resonance, even though the body is in the state of rest or semi-activity, the blood is still moving with great vigor.*

In traditional Chinese medicine, there is a common symptomology referred to as "blood stagnation". Blood stagnation accompanies many forms of illness and is simply understood to mean that "the blood is not flowing" (as it should). A key tenet of Chinese medicine it that: "Qi (vital energy) moves the blood". Consequently, it is generally thought that "blood stagnation" is rooted in "qi stagnation". Breathing is assumed to be the main source of qi and the motive force behind its movement – the Chinese characters for "qi" and "air" are the same! This makes perfect sense: *Breathing moves the blood.*

Limit the action of breathing and the role of generating and maintaining blood pressure falls to the heart and cardiovascular system – with serious consequence. In summary, I offer this argument:

> *A primary function of heart rate variability is moderation of arterial pressure, where it works in opposition to the breathing induced respiratory arterial pressure wave to regulate increases and decreases in dynamic arterial pressure. Relative to this function, resonance may be characterized as the moment when the heart rate variability cycle is performing this service to the maximal extent. Consequently, it also defines the productive range of respiration.*

4

Coherence

In *The Rainbow and the Worm*, Mae-Wan Ho uses the term "coherence" as an expression of "wholeness" of the living system, "incoherence", or deviation away from coherence, being a measure of its entropy. "A coherent system is totally transparent to itself, as all parts of the system are in complete, instantaneous communication."[9] In neurofeedback circles, "coherence" is used to express psycho-physiological balance, a meaning that is very consistent with "wholeness". As applied in physics, "coherence" also means physical *congruity* or *consistency,* in this case of the heart rate variability wave or cycle. As we know that heart rate variability is in effect the signature of the autonomic nervous system, and that the autonomic nervous system is the operations and maintenance substrate, "coherence" is a measure of the *balance, communication, and unity* of the human organism.

In the West, we tend to think of the mind and body as being separate. In his work, Shifu John Painter (9 Dragon Baguazhang) speaks of the relationship between mind and body as the "psycho/soma mobius." The fascinating thing about the mobius is that while you initially perceive it as *obviously* having two sides, if you examine it carefully it only has one!

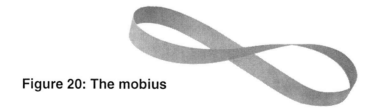

Figure 20: The mobius

The New Science of Breath

The mind/body mobius concept is very consistent with yogic theory as well as with the latest scientific findings that where the brain was once considered the residence of the mind, the mind is in fact distributed within every cell of the body, potentially extending even beyond the body's accepted physical limits. The quantum understanding is that "body is mind" and "mind is body", i.e. they are literally "one". Not surprisingly, heart rate variability is also reflective of both body and mind, hence the psycho-physiological correlates of autonomic nervous system balance, a few of which are cited in the table below.

Imbalance	Correlate	Balance
tense	physical tension	relaxed
discomfort	physical comfort	comfort
anxious	mind state	at ease
muddled	thinking	clear
extreme	sensibilities	tempered
impeded	mind-body communication	free flowing
weak	short term memory	strong
reduced	sphere of awareness	expanded
reduced	intuition	increased
reduced	openness to new ideas	increased
defensive	interpersonal communication	accepting
acid	serum pH	balanced
higher	average heartbeat rate	lower
much higher	heart duty cycle (work)	much lower
higher	blood pressure	lower
lower	heart rate variability amplitude	higher
lower	gas exchange efficiency	higher
decreased	breathing depth	increased
higher	breathing frequency	lower

Figure 21: Psycho-physiological correlates of autonomic nervous system balance

You may recall the prior assertion that "the autonomic nervous system alone does not possess the ability to govern without our conscious participation". Here I would like to generalize this further to say that **the body** *alone does not possess the ability to govern itself without our conscious participation*. If we accept for a moment the quantum view that the mind is the body, and the body is the mind, by not participating consciously in the functioning of the body we are in fact splintered, our mind not participating in our body's réalité. We are out of communication with our selves, incoherent, and without unity.

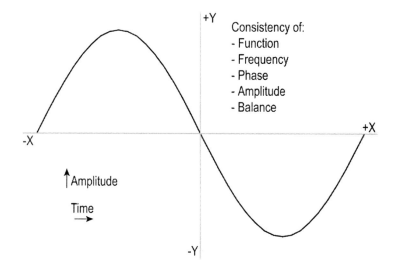

Figure 22: A pure sinewave

A theoretically pure sinewave, as depicted in Figure 22 above, is a wave function with perfect coherence, perfect consistency of amplitude, frequency, phase, and spectral purity, and perfectly balanced around an X axis with a Y coordinate of zero. The oscillator that produces a perfect sinewave is in itself perfectly coherent, the perfect sinewave being an expression of its coherence. By the same token, a perfectly coherent heart rate variability pattern is perfectly sinusoidal. And, the living system that produces it is perfectly coherent (recognizing that the human organism is a nonlinear dynamical system).

Consequently, the heart rate variability signature we aspire to during resting homeostasis is perfectly sinusoidal and centered around a Y coordinate of zero, the X-axis being the conceptual dividing line between sympathetic and parasympathetic emphasis. A heart rate variability cycle which approximates to this description exists at only one moment, the moment of perfect resonance. This moment represents ideal homeostasis, the moment of perfect balance/communication and minimal entropy/chaos.

Although reasonable, it is not intuitively obvious that the rate at which the heartbeat changes is in fact an *intrinsic* autonomic nervous system rhythm, nor that it is the outcome of a complex process that is mediated by the autonomic nervous system. While the medulla oblongata is known to generate rhythmic potentials that stimulate respiratory muscle groups, the relationship of that function to heart rate variability is at present unclear. In any case, the rhythm of resonance is detectable throughout the body. In *The Heartmath Solution*, the authors demonstrate that during the state of cardiopulmonary resonance the HRV rhythm can clearly be seen to modulate brainwaves as measured by an electroencephalograph![10] Actually, the same is true of an incoherent HRV rhythm, i.e. it also modulates the brainwaves, but because it is incoherent it is difficult to discern the pattern. The HRV rhythm can also clearly be seen to modulate the sympathetic nervous system as detected in the hands via galvanic skin response.

Referring once again to Figure 14, the primary factor affecting coherence, overall integrity and consistency of the HRV rhythm, is synchrony or lack thereof between the oscillator's resonant frequency, which aspires to 1 cycle in ~12 seconds while erect and in a state of rest or semi-activity, and the "respiratory signal". These two rhythms are in a sense additive. If they are in phase, the result is maximal HRV amplitude and maximal coherence. If they are 180 degrees out of phase, the result is, in theory, an HRV amplitude of zero – in practice, an incoherent HRV with greatly diminished amplitude. Because the majority of people breathe both relatively fast and relatively asynchronously, their oscillator and breathing rhythms tend to be highly asynchronous, resulting in a heart rate variability signature with irregular amplitude, phase, and frequency, i.e. incoherence.

5
Coherent Breathing

Breathing is the primary means by which we participate in the felt but unseen work of the autonomic nervous system. The physiological mechanism by which this occurs is the "breathing bridge", the breathing link between somatic and autonomic nervous systems.

There is one heart rate variability frequency and one breathing frequency, again depending on circumstances, that equals optimal homeostasis, ideal sympathetic/parasympathetic balance, and wholeness. This is the frequency of resonance. When one breathes at this frequency with appropriate depth, a few things happen. One, the autonomic nervous system centers itself about the dividing line between sympathetic and parasympathetic balance, such that when you inhale the autonomic nervous system swings toward sympathetic emphasis and as you exhale it swings toward parasympathetic emphasis, all the while maintaining its centeredness about the sympathetic/parasympathetic dividing line. (This conceptual dividing line is the average heartbeat rate.) This is the approximate 12-second *Fundamental Quiescent Rhythm,* consisting of a sympathetic phase and a parasympathetic phase which are of equal amplitude and opposite phase resulting in net balance.

In reality, it is far from being this simple. Every bodily action, including exhalation, stimulates sympathetic action to some degree. During inhalation there is sympathetic action with minimal parasympathetic emphasis, and during exhalation there is sympathetic action with maximal parasympathetic emphasis.

The New Science of Breath

Resonance defines the *neutral state* – in this state, the sympathetic bias we are used to living with disappears. With it goes muscle tension, anxiety, mental chatter, and discomfort. These feelings are replaced with the feelings of ease, peace of mind, stillness, comfort, and an overall sense of *harmony* and well-being.

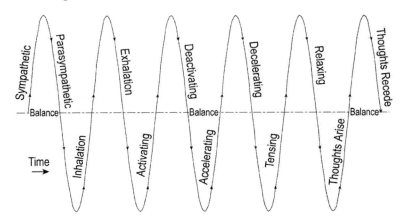

Figure 23: Fundamental quiescent rhythm and sympathetic/parasympathetic correlates

Cardiopulmonary resonance can be elicited either consciously or unconsciously. Some promote the self-discovery of autonomic balance and resonance via the cultivation of positive emotions and a felt sense of harmony and appreciation. In this way, heart and breathing rhythms "find" each other, resulting in autonomic balance and resonance. The practicer learns to associate the state of resonance with these positive feelings. It is my understanding that autonomic balance and cardiopulmonary resonance may also be elicited via hypnotherapy, although I have not seen it demonstrated.

Alternatively, *The New Science of Breath* promotes accessing the state of autonomic balance and cardiopulmonary resonance via conscious breathing. The practicer may bring positive emotions of their choice to the practice if they wish, but it is not required. The state of balance will elicit balanced emotion. You can think of the differences in these two approaches as equating to the essential yogic alternatives of "cultivating the body via cultivation of the mind" vs. "cultivating the mind via cultivation of the body". Much more will be said about Coherent Breathing and yoga in chapter 7. For now, you may think of Coherent Breathing as a specific form of "pranayama".

There are two ways to elicit autonomic balance and cardiopulmonary resonance via breathing. The first method is to monitor the heart rate variability rhythm and consciously synchronize the breathing rhythm with the heart rate variability rhythm. To do this, you synchronize your inhalation with positive-going heartbeat rate and your exhalation with negative-going heartbeat rate, changing from inhalation to exhalation at heart rate peaks, and exhalation to inhalation at heart rate valleys. This is a very effective way to learn as well as practice Coherent Breathing. The method requires monitoring the heart rhythm, usually with an instrument. I have used an HRV monitor and an electronic stethoscope as well as other lab instruments for this purpose.

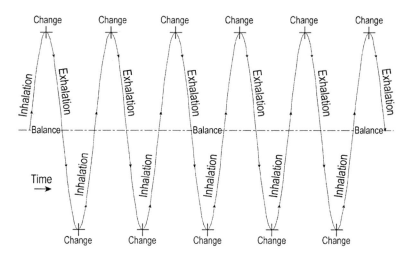

Figure 24: Inhalation/exhalation and changes at peaks and valleys

The second method is to breathe at the frequency of resonance which is .085 cycles per second or 5 cycles in 1 minute (58.8 seconds to be exact), this frequency being essentially the same for adults while in an erect body position and otherwise at rest or semi-activity. Because the autonomic nervous system understands this frequency to be that of balance and ideal homeostasis, the heart rate variability rhythm will change to align with this breathing rhythm, even if at the moment it is very different. For this reason, even a few minutes of breathing at the *frequency* of resonance can move a person from being tense and anxious to being calm and relaxed – just that quickly.

So far, our discussion has focused primarily on breathing at the frequency of resonance, what *The New Science of Breath* refers to as the *Fundamental Quiescent Rhythm*. While breathing frequency and depth are two sides of the same coin, it is possible to breathe slowly yet shallowly. Therefore, it is possible to breathe at the frequency of resonance with any depth, as long as gas exchange is adequate and the autonomic nervous system does not decide to step in to get more airflow going. However, the opposite is not true – it is not possible to breathe rapidly yet deeply. This is simply because rapid breathing does not allow time enough to inhale and exhale to any significant degree. Therefore, by definition, as breathing frequency increases, depth must decrease.

Depth is the critical factor in achieving HRV amplitude. This is because inhalation stimulates sympathetic emphasis corresponding to HRV peaks, and exhalation stimulates parasympathetic emphasis corresponding to HRV valleys. If the extent of inhalation and exhalation is relatively low, the deviation in sympathetic and parasympathetic emphasis will be relatively small. Alternatively, if the extent of inhalation and exhalation is relatively great, the deviation in sympathetic and parasympathetic emphasis will be relatively large. An example of this can be seen in Figure 25 below, which depicts breathing at the rate of 5 cycles per minute with shallow depth, immediately followed by deeper breathing. As can be seen in this particular case, shallow breathing yields an HRV amplitude of 17 beats, followed by deeper breathing yielding an HRV amplitude of 40 beats.

Figure 25 is a visual example of respiratory sinus arrhythmia (RSA) in action, i.e. it clearly demonstrates how heartbeat rate varies as a function of breathing frequency and depth. It is important to note that the HRV "ceiling" does not vary as a function of breathing depth – it is a function of *real time sympathetic emphasis*, this being dictated by breathing frequency, time of day (i.e. biorhythm), body inclination, digestive status, etc. Alternatively, the HRV "floor" is very much a function of depth. As HRV amplitude is the delta between *ceiling* and *floor*, depth determines HRV amplitude.

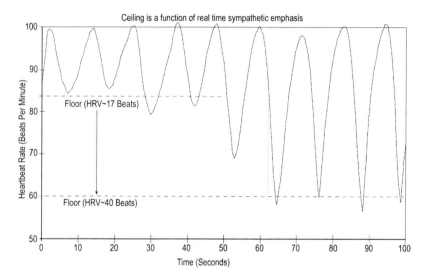

Figure 25: Shallow breathing immediately followed by deeper breathing (both at 5 cycles per minute)

While the peak of inhalation, the moment when inhalation ceases and exhalation begins, *aligns* with but does not determine peak heartbeat rate, the peak of exhalation, the moment when exhalation ceases and inhalation begins, very much influences the valley heartbeat rate. In other words, the sympathetic ceiling is in effect the *baseline* – it does not vary with breathing depth; the floor represents deviation from the baseline, this deviation principally being a function of exhalation.

Breathing frequency is the rate of inhalation and exhalation. What exactly is *depth*? We tend to think of depth as "fullness of the lungs", "more air". *Internally*, the definition is quite different.

The New Science of Breath

Per this theory, we can see that heart rate variability is inextricably linked to the mechanical action of breathing where it serves to regulate the respiratory arterial pressure wave. (Please refer to Appendix A, Figures 37-39 for this discussion.) It accomplishes this via the baroreceptor reflex. The baroreceptor reflex is a nervous function that is distributed throughout major arteries, its role being that of maintaining peripheral arterial pressure within viable limits. During cardiopulmonary resonance, when arterial pressure rises, in this case as a consequence of exhalation, i.e. increasing positive thoracic pressure and consequent increased blood flow to the heart, the baroreceptor communicates this increase to the autonomic nervous system. The autonomic nervous system throttles rising pressure by slowing heart rate and relaxing (enlarging) arteries. This moment corresponds to the HRV peak.

Upon inhalation, diaphragm flexion occurs resulting in strong negative thoracic pressure (a vacuum) and consequent expansion of blood vessels in the chest, storing blood, and reducing flow to the heart for the period of inhalation. This results in a decrease in peripheral arterial pressure. The baroreceptor reflex communicates this decrease in pressure to the autonomic nervous system, which responds by increasing heart rate and constricting arteries, thereby maintaining peripheral pressure within viable limits. This moment corresponds to the HRV valley.

Therefore, from the autonomic nervous system perspective, depth is a function of the range of diaphragmatic action – resulting in positive and negative thoracic pressure – resulting in arterial pressure wave peaks and valleys – resulting in autonomic nervous system down/up regulation – resulting in heart rate variability peaks and valleys, respectively, thus forming a closed-loop feedback system that maintains arterial pressure within viable limits.

Early studies of heart rate variability held that HRV amplitude was governed by compliance of heart tissue, "compliance" being an indication of one's true "biological age". While there may be some truth to this, we are coming to understand that heart rate variability amplitude is fundamentally determined by a) the frequency and depth of breathing, and b) relaxation.

Breathing *frequency* strongly influences sympathetic function. For this reason more rapid breathing results in stronger sympathetic emphasis. Breathing depth strongly influences parasympathetic function, and for this reason, deeper breathing results in stronger parasympathetic emphasis. Consequently, there is a breathing frequency and depth where sympathetic and parasympathetic effects are equal, thus resulting in autonomic nervous system balance. This relationship can be seen clearly in Figure 26 below. (Please note that Figure 26 assumes underlying sympathetic bias, i.e. strong parasympathetic stimulation plus weak sympathetic stimulation result in net balance.)

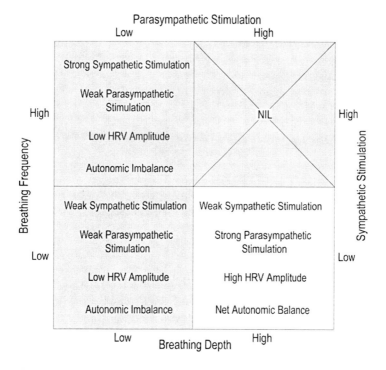

Figure 26: 4-quadrant depiction of the relationship between frequency, depth, and autonomic emphasis

Nevertheless, numerous studies support the fact that HRV amplitude (in the absence of intervention) does decline with stress and age. Why? What is going on? As we have seen, HRV amplitude is indicative of the degree to which the mechanical action of breathing is, in effect, offloading the heart and cardiovascular system

The New Science of Breath

of the responsibility for generating and maintaining arterial pressure. Secondly, maximal HRV amplitude correlates with a *respiratory arterial pressure wave* of maximal amplitude. Per Figure 27, the arterial pressure wave governs the rate and efficiency at which oxygenated blood is transported to the extremities and the rate and efficiency at which carbon dioxide laden blood is returned to the heart and lungs, i.e. "circulation".

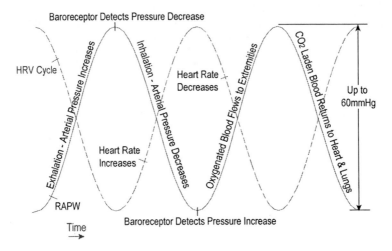

Figure 27: Respiratory Arterial Pressure Wave and HRV cycle at cardiopulmonary resonance

With this in mind, compare the robustness of cardiopulmonary operation during deep resonant breathing characterized by Figure 8 with that of insufficient breathing, Figure 7. One can see from this how and why "blood stagnation" can occur as a consequence of very shallow asynchronous breathing.

It is not news that breathing fuels energy production, energy production being a function of the reaction of oxygen with nutritional components. While research into this area is far from complete, there is clear indication that deep resonant breathing, as evidenced by robust HRV amplitude with a high degree of coherence, elevates "bioenergy". This increase can be observed, particularly during extended periods of breathing combined with relaxation and stillness, where it is thought that optimal breathing contributes to bioenergy generation and stillness/relaxation minimizes the "load" or usage thereof. Over time, this results in a gradual increase in measurable biocurrent, this current being a function of either an increase in biopotential or a decrease in body impedance, or both.

Under exacting conditions, this current flow can be observed and measured. Figure 28, below, depicts the rise in *current* between the body and earth, observed for a period of 30 minutes while employing Coherent Breathing, relaxation, and stillness. As can be seen, alternating current increases from an average of approximately 16 microamps with peaks as high as 19 microamps, to an average which approximates to 18 microamps with peaks as high as 21 microamps. While this change appears rather small, in terms of biological potential it is quite high. A little later, this rise in bioenergy will be shown to be instrumental in facilitating the yogic phenomenon known as the "kriya".

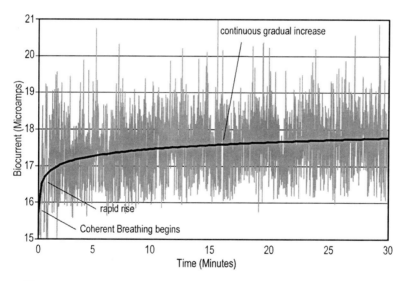

Figure 28: Increase in biocurrent as a result of Coherent Breathing, relaxation, and stillness

What is the physiological source of this elevated biocurrent? Does it result from sustained autonomic nervous system balance? Is it a function of oxygenization combined with minimizing the energetic burden? If one breathes optimally on an ongoing basis, is relatively higher bioenergy a homeostatic norm?

The New Science of Breath

If we look at the rise and fall of human physiological capacity vs. age, including peak athletic performance, hormone production, resting blood oxygen level, working blood oxygen level, and cellular energy production, a consistent graph emerges, all of these factors exhibiting essentially the same shape and slope, that of the bold line of Figure 29.

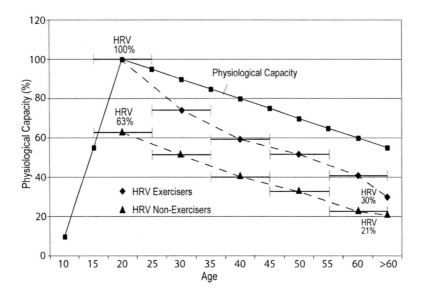

Figure 29: The rise and fall of human "capacity" vs. age

As can be seen, "capacity" rises through approximately age 20, at which point it begins a steady, almost linear, decline. Interestingly, Helmuth Frank's s*ubjektives zeitquant* or *subjective time quanta*, the rate at which the conscious mind is able to perceive individual events in time, in effect a "consciousness bandwidth" metric, follows this same basic curve,[11] making it roughly representative of both physiological and perceptual competence.

Citing data from a 1992 study by R. De Meersman as presented in *American Heart Journal*, two HRV curves are superimposed, the upper curve being associated with exercisers and the lower curve with non-exercisers. Participants were divided into 6 age groups ranging from 15-25, 25-35, 35-45, 45-55, 55-65, and over 65. The HRV of exercising 15-25 year-olds, being the highest, equates to 100% of HRV capacity. The curve associated with exercisers declines by 70% over the period and the curve associated with non-exercisers declines by 42% over the period. (Note that the HRV of exercisers starts out almost 40% higher and ends almost 10% higher than non-exercisers.)[12] De Meersman's study is particularly relevant because, during his measurement of HRV amplitude, participants were assisted in breathing at the rate of 6 breaths per minute, thereby exhibiting HRV amplitude which approximates to that of resonance. As can be seen, HRV amplitude, being an accurate indicator of "biological age", tracks very closely with overall "capacity", their slopes being very similar.

Yet, in our experience, HRV amplitude is far from being an absolute and responds dramatically to cultivation via breathing. This leads to the question, how much of the decline in HRV with age is a consequence of "normal" aging vs. pathological senescence, specifically that resulting from sustained suboptimal breathing and related autonomic imbalance? Does the body's bioenergy/biocurrent follow this same curve? Might a decline in bioenergy due to suboptimal breathing precipitate the decline of other factors? In yogic circles, breathing is known to "expand" consciousness – does breathing affect the *bandwidth* of consciousness per Frank's *subjective time quanta*? Interesting questions yet to be answered…

In the context of COHERENCE, a number of customers who have purchased Breathing Pacemaker® therapeutic audio products have gone on to acquire heart rate variability biofeedback instruments for their personal use. It is not unusual that, at the outset, their HRV amplitude is relatively low. In other words, even though they are breathing at the *Fundamental Quiescent Rhythm* their HRV amplitude may be 10 beats vs. 20, 30, 40, or even 50 beats.

I have worked with a number of them and find that, in general, it is not because their peak amplitude is high but because their valley amplitude is not low. Recalling the discussion of HRV ceiling and floor, while HRV amplitude is known to decline with age, both *breathing depth* and *relaxation* play a key part in moderating amplitude relative to its biological potential. Therefore, progress can usually be facilitated by encouraging people to relax deeply upon exhalation, specifically focusing on relaxing the bridges, including the eyes, the jaw, the perineum or pelvic floor, the hands and the feet. Usually within a week or two, they've added 10 points to their HRV amplitude and, for those that stay at it, another 10 beats within a month or so. Therefore, while it does take time to cultivate robust amplitude, especially if you are not practicing another physical art, i.e. yoga, martial art, sport, dance, etc., that has required you to cultivate relaxation, in most cases it is *very* possible. This fact will be clearly demonstrated in the following chapter.

So, what of HRV amplitude and autonomic "tone" and "responsiveness"? The central nervous system will limit range of motion to that which is used. For this reason, if we were to deliberately walk with 3-inch steps for a month, by the end of that month, it would be difficult to take a step longer than 3 inches. This is the same mechanism that results in loss of range of motion during extended periods of immobility, for example, immobilization following a serious accident. By the same token, *range* of heart rate variability is built and maintained through use. What you don't use, you lose. The central nervous system literally takes it away. The way you "use" heart rate variability range is by breathing. This is the reason that heart rate variability responds dramatically to aerobic exercise, i.e. it necessitates deeper synchronous breathing, regular periods of deeper synchronous breathing building both amplitude and coherence.

As previously stated, heart rate variability amplitude is an accurate indicator of biological age and correlates highly with *all causes of mortality.* Yet, HRV amplitude is highly modifiable via breathing, since breathing frequency and depth to a large extent determine HRV amplitude, at least within the scope of one's *true* biological age. *If this is true, then it is breathing sufficiency that correlates highly with all causes of mortality.* Is this surprising?

A foundational concept of Taoist yoga is that "Man is born with the breath anchored in the abdomen. As man ages, the anchor diminishes and the breath floats upward into the chest, the throat, and finally exits through the mouth, thus resulting in death." Taoist yoga admonishes one to struggle against this tendency by "keeping the qi and the mind in the dan tien", a key aspect of this practice being slow, deep, mindful breathing.

I close this chapter with these beliefs. Mankind has known since the dawn of time that breathing is necessary for health, well-being, and longevity. Yet, it hasn't been clear *why* breathing is such a critical factor. Nor has it been clear exactly *how* to breathe to maximize benefit.

However, we now have a quantitative measure of breathing effectiveness: the heart rate variability cycle. While its relationship to breathing, as herein described, has yet to be embraced by the medical community at large, robustness of the HRV cycle is accepted as an accurate measure of *mortality risk from all causes*. As the HRV cycle is indicative of autonomic status it is understandable that it influences all aspects of life, mind and body, conscious and subconscious.

The heart rate variability cycle is a mirror image of the breathing induced respiratory arterial pressure wave, a primary function being that of regulating instantaneous arterial pressure. As such, it also mirrors the degree to which breathing is moving blood in the body, or possibly more concerning, the degree to which it isn't.

Coherent Breathing elevates blood flow, bioenergetic potential, and current flow within the human organism. As life is energy, elevated bioenergy enhances all aspects of life including health, healing, mental function, spiritual experience, and performance. Again, this notion is very consistent with age-old Eastern wisdom and understanding.

Note: "Chi" and "qi" refer to "the vital essence" in air as well as in the body, the latter being derived principally from breathing. The "dan tien", commonly translated as "elixir field", is a zone located slightly below the navel and inside the abdomen. It is considered the "sea of qi" in the body. It is also referred to as the "yellow origin" signifying its role in the cultivation of the alchemical "golden flower".

6

The Biometrics of Balance
Case Observations Employing Coherent Breathing

by Dee Edmonson

From our bodily user perspective, breathing has two primary attributes: frequency and depth. That is, these are the aspects of breathing that we can consciously control.

The frequency and depth that is appropriate for one circumstance is not appropriate for others. For example, the frequency and depth of Lamaze breathing for childbirth is not appropriate during your morning walk. Likewise, the frequency and depth of fight-or-flight is not appropriate in the absence of threatening circumstance. Yet, this is what I see in many clients.

Since the introduction of the RESPIRE 1 Breathing Pacemaker CD and Coherent Breathing in early 2005, I have used it with well over 100 clients and family members ranging in age from 6 to 60. Diagnosis or presenting conditions included, but were not limited to, generalized and performance anxiety, depression, drug and alcohol addiction, sleep dysfunction, mood disorders, attention deficit disorder (ADD and ADHD), traumatic brain injury, tic disorder, obsessive compulsive disorder (OCD), and hypertension.

I was able to assess the heart rate variability and breathing patterns of many of them, and generally, where their HRV demonstrated low amplitude or coherence, it was accompanied by short, shallow breathing. (I observed their frequency and depth of respiration visually and, on occasion, by using a stethoscope to assess depth, flow, and rhythm.) Short, shallow breathing was also frequently accompanied by elevated high beta at multiple EEG sites and a relatively high electro-dermal response (skin conductivity), both of which can be strong indicators of stress and anxiety. Interestingly, some clients expressed an awareness of their stress and anxiety while others only began to understand that their internal or external symptoms were stress and anxiety related when provided with the results and interpretation of their assessment.

I began introducing the practice of Coherent Breathing to clients who exhibited relatively low or incoherent heart rate variability patterns, and found that they responded promptly and very favorably, often making significant gains in HRV amplitude and/or coherence in a single session. Over the course of months, as I introduced Coherent Breathing to a wider base of clients, a clear pattern emerged.

First, when asked how they felt after 8 minutes or more of Coherent Breathing, almost without fail the response from clients was: "I feel different", "I feel peaceful", "I feel calm", "I feel better"... And, they often left the center reporting that they felt brighter and more optimistic.

Second, I could clearly see that Coherent Breathing not only affects heart rate variability, but affects other biometrics as well. For example, elevation in hand temperature, reduction in electro-dermal response or skin conductivity, reduction in high frequency beta brainwaves, and an increase in alpha brainwaves where relevant. Clearly, Coherent Breathing was having an impact on many levels.

As time progressed, I continued to see this with increasing numbers of clients. Desiring additional input, I contacted several of my close colleagues in the neurotherapy field and discussed with them the results that I was seeing from the use of Coherent Breathing and RESPIRE 1. One such colleague is Dr. Elsa Baehr, Clinical Associate, Department of Psychiatry & Behavioral Sciences, Northwestern University Medical School, and Baehr & Baehr LTD., well noted for her research and contribution in the application of neurotherapy to the treatment of depression. Another is Dr. Roger Riss, Department of Neuropsychology at Madonna Rehabilitation Hospital, in Lincoln, Nebraska. Roger's primary focus is the application of neurotherapy to traumatic brain injury, acute stress and anxiety disorders related to motor vehicle accidents, and resulting post traumatic stress disorders. I provided Elsa and Roger with copies of RESPIRE 1 and explained both its use and the session protocol that I was employing. Their observations have been consistent with my own. That is, the application of Coherent Breathing tends to facilitate "desired" biometrics, those biometrics that have been identified over time to equate with well-being.

The following case observations are provided to demonstrate the effects of Coherent Breathing. Please note that Coherent Breathing was applied incrementally as but one aspect of a comprehensive program that I offer clients. I employ many instruments and programs, each of which has its respective place depending on individual needs, including EEG neurotherapy, heart rate variability (HRV) training, thermal and electro-dermal response (EDR) biofeedback.

Prior to any client treatment intervention, a comprehensive intake assessment is completed, including presenting problems, psychological/social, medical history, medications, nutritional habits, family observations/concerns, and in the case of children, school performance and behavior. If indicated, a Neuro-Physiologic Assessment (NPA) is done using EEG biofeedback techniques. A Test of Variables of Attention (TOVA) is also included if warranted. When all evaluations are complete, I review the information with the client and/or family members, initial goals are established, and a program is outlined defining treatment methods and session frequency.

The following observations are made within the context of structured sessions as opposed to a rigorous scientific study. However, anticipating that a number of the case observations might be cited in this work, I was careful to maintain clean and clear guidelines. An HRV training program that recognizes and readily facilitates heart and breath synchrony was employed, the same HRV program being used to assess autonomic status baselines as well as for training and for assessing progress. With few exceptions, HRV baselines were established prior to initiating breath training. Coherent Breathing supported by RESPIRE 1 was the only breathing method used other than the participant's normal breathing pattern. With the exception of one, none of the participants had previous breath training.

This information is being offered for educational purposes. No claims regarding specific health benefits are being made. The information is not intended to replace the services of a medical or mental health professional nor to diagnose, treat, cure, or prevent any disease.

In the following discussion, language and terminology common to the field of neurotherapy is often used. An explanation of these terms is provided in Appendix A. All of these observations occurred during 2005. Names have been changed to ensure the client's privacy.

Case Observations:

1. Maria *(anxiety)* is a bright, energetic 19 year-old whose primary complaint was poor test performance in school and consequent low grades. She is very active in school sports including track, swimming, and lacrosse. Maria believed her low grades to be due to not studying hard enough. Her parents believed that she might benefit academically from EEG biofeedback training. An NPA was done as well as a TOVA. The results were indicative of anxiety, and specifically performance anxiety relating to cognitive tasks.

Maria's initial HRV exhibited inconsistent amplitude averaging 10-20 beats with 51% coherence. Her initial electro-dermal response (EDR) was quite elevated, averaging 19-20 micromhos on an effective scale of 0-30, this being indicative of relatively strong sympathetic emphasis. Her treatment program would include heart rate variability (HRV) training, EEG and electro-dermal response (EDR) biofeedback, as well as the cultivation of stress management and coping skills.

Maria had a very short window of time, requiring an intense program of 2 sessions per day for 5 days before leaving on vacation for 3 weeks. Session number 1 was scheduled for later that day and started with HRV training and EDR biofeedback. During this first session her HRV continued to demonstrate inconsistent amplitude and coherence, and she expressed that she was "quite bored".

With EDR biofeedback, Maria could watch a graph on a computer monitor positioned conveniently in front of her. The objective was to begin lowering her EDR from the 18-20 range with eyes open. This proved to be very challenging and unproductive, therefore I instructed Maria in the use of Coherent Breathing and asked her to close her eyes and breathe in synchrony with "Vocal Instructive Sequence" from the RESPIRE 1 CD. I observed her respiration, providing coaching, until it was in synchrony with the rhythm on the CD. I then switched tracks to "Clock & Bell" and continued to observe her breathing as well as her EDR. Session 1 ended with a decrease in her EDR of 10 points.

The next morning Maria arrived for session 2. Like session 1, the session consisted of Coherent Breathing, HRV training, and EDR biofeedback. Her initial EDR again measured 20 with eyes either open or closed, and again, her HRV demonstrated amplitude of 10-20 beats with ~50% coherence. She was not very enthusiastic and noted that she "really wanted to be somewhere else". Again I employed Coherent Breathing during EDR biofeedback and she was again able to effect a reduction of about 10 points in her EDR by the end of session 2. This basic protocol of Coherent Breathing, HRV training, and EDR biofeedback continued for the next two sessions.

In the fourth session, I incorporated Coherent Breathing in conjunction with HRV training. In this 20-minute training segment, she demonstrated pronounced changes, averaging 28 beats with 92% coherence. At the end she noted that she felt "peaceful". I continued with the use of Coherent Breathing during EDR biofeedback and by session 8 her EDR with eyes open had decreased to 6-7.

After vacation, Maria returned to continue a more focused program including EEG biofeedback. The same protocol was continued as noted previously. In this tenth session, her HRV amplitude was 25 beats with 97% coherence and her EDR measured 10 declining to 6 over the course of the session, with eyes open. Maria was able to participate in 8 more sessions prior to returning to school, each session consisting of Coherent Breathing and HRV training, followed by EDR and EEG biofeedback. At the completion of her final session Maria's HRV amplitude was 28-30 beats with 97% and her EDR was 5-6. Maria returned to school with two copies of the RESPIRE 1 CD, one for herself and one as a gift for her roommate who "had difficulty sleeping".

2. Joseph *(hypertension, pain, and anxiety)* is a 49 year-old master carpenter and home builder. He and his wife contacted me in late February of 2005 stating that Joseph was having difficulty regulating his blood pressure, which at that time was 180/120. He was on an antihypertensive medication and a low sodium diet. He did not smoke, and his caffeine intake was limited to 1, perhaps 2 cups of coffee per day. He was getting a moderate amount of exercise and was able to work several days a week remodeling homes.

At times Joseph was anxious and irritable and often felt very fatigued with sudden drops in energy. Joseph stated that he had had chronic pain as the result of 39 broken bones, including his hip and spine, from years in construction work. He also related that, while he was presently free of alcohol use, he had had an alcohol addiction. Years ago he had turned to alcohol as a way to deal with physical pain and feelings of anxiousness. He was seeking alternative means to achieve relaxation and moderate his anxiety. EEG biofeedback was not a practical option for him, and his prior attempts to use heart rate variability (HRV) training were met with frustration. In March of 2005, I provided Joseph with a copy of RESPIRE 1 and instructed him in its use. Thereafter I saw him every two weeks. In between he gave me weekly updates on his condition and kept me informed of the results of his visits to his doctor and his use of medications.

In October, I asked Joseph if he would like to relate his story for inclusion in this book. Here are his comments:

"The change I have noted with Coherent Breathing is the ability for me to achieve a calm state of mind and inner peace. I've been trying to achieve this for a long time – feeling comfortable in my own body. Even after decades of sobriety, I still felt uncomfortable within myself.

The New Science of Breath

"In March, I started using RESPIRE 1. I often used it with headphones in order to find a place of quiet. At first I felt lightheaded and dizzy, but with practice learned to do the breathing with ease. It started to give me what I wanted, a feeling of true calm. I progressed to being able to use the skill for 5-10 minutes and was able to carry the rhythm in my head and to visualize the sinewave. When I did not have the CD, I learned to use my watch to time my inhalation and exhalation.

"For me it is a way to maintain my sobriety. The way in which anyone with an addiction to alcohol handles stress determines his ability to maintain his sobriety. Now I can get to a place of calm. My blood pressure has dropped to 127/78. And, with the doctor monitoring me, I'm now on half the dose (of antihypertensive medication) every other day. My heart rate has decreased, my mind has stopped racing, and my sleep has improved. I'm back to living better. Even my friends got nicer."

Joseph went on to relate that had a fear of the dentist. Just the anticipation "was an easy trip to anxiety". This summer he had to have oral surgery and declined the use of anesthesia other than a local injection. Moments before surgery, he sat in the chair and used his watch to practice Coherent Breathing. He was able to "achieve the state of calm for the entire procedure" and "used breathing after the surgery to manage pain".

Joseph also had to have trigger point injections to deal with the pain he was experiencing from construction injuries. He related that prior to the injections his blood pressure was recorded at 180/100. By the time he was assessed by the physician just moments before the injections, by using Coherent Breathing he had lowered his blood pressure to 127/78.

Joseph continues to use Coherent Breathing daily and is successful in maintaining his blood pressure in the range of 128/78. He expresses that his life is more "peaceful".

3. Steve *(anger)*, 38, is married with two children. He is employed as a detention officer at a large correctional facility in north Texas. At the time of his assessment, he related that "my life is in a spin and headed in the direction of mayhem". He referred to himself as the "Tasmanian Devil". He was suffering from seemingly uncontrollable outbursts of anger. These outbursts were jeopardizing his job and his family. He was also experiencing feelings of anxiety and was having difficulty sleeping and concentrating. Steve stated that he smoked, and noted that the number of cigarettes increased with his stress level.

During the initial intake, he found it difficult to discuss his anger and feelings of frustration, but with the loving support of his wife he began to feel more at ease. He related that his anger manifested itself in hitting walls, yelling, shouting, and lashing out at people, both on the job and at home. This pattern of behavior had been ongoing since the age of 10. Only during his military service between the ages of 18 and 20 did he note that his tendency to anger abated. He also noted that he felt much more at ease when working outdoors with a greater flexibility of work schedule than his present job affords him.

Steve's NPA indicated elevated high beta (19-33 Hz.) in frontal, central, and occipital areas, and increased frontal delta/theta (2-7 Hz.). Steve's initial heart rate variability amplitude was low, 5 beats, with low coherence. His breathing pattern was short and shallow, involving his upper chest only. A program of EEG biofeedback was established and goals were set with both Steve and his wife. This involved 2 sessions per week for several months.

Steve's training program began within days of his initial assessment, with an initial focus on EEG biofeedback. At his first session, I instructed him in the application of Coherent Breathing and provided a copy of RESPIRE 1 for use at home. Over the following weeks, changes were noted in Steve's EEG pattern, specifically the desired decreases in frontal high beta and delta/theta frequencies. He began to report clearer thinking and a better ability to focus, but he had not experienced much improvement in his ability to control his anger. His HRV amplitude and coherence had improved only slightly, despite frequent coaching.

I assessed Steve's comprehension of Coherent Breathing and use of the RESPIRE 1 CD, and it was clear that he did not fully understand how to use it, nor was he practicing the method at home. With more in-session training and practice at home, his HRV amplitude and coherence gradually increased. Over the course of several months, his HRV amplitude climbed from 5 to

The New Science of Breath

10, to 15, and from 15 to its present 18-20 beats with high coherence. During this process, Steve began to experience notable changes in his ability to control his behavior. (Figures 7 & 9 are representative of low HRV amplitude with low coherence and moderate HRV amplitude with high coherence, respectively.)

Steve's EEG biofeedback included training of both fontal and occipital areas, and he practiced Coherent Breathing and HRV training prior to each session. For his last frontal training session, Steve elected to use Coherent Breathing (with eyes open). With the use of Coherent Breathing, he and I noted a sudden decrease in the amplitude of both delta/theta and high beta components. With this interesting development, I elected to employ the "Clock & Bell" track during much of Steve's alpha/theta training. With eyes closed, without RESPIRE 1, there was only a moderate increase in alpha amplitude. As Steve began RESPIRE 1-guided Coherent Breathing, his alpha amplitude increased significantly. During this session Steve went on to access a deeply relaxed state with vivid imagery and experienced a "very new feeling", the feeling of being totally relaxed.

After five months of training twice per week, a smiling, humorous, brighter, Steve is emerging. These are his words:

"I've noticed an increase in my ability to focus. I can change the outcome of a situation rather than the situation changing me. There is a whole lot less anger (now). I'm not quick to judge the behavior of people (confined to the detention center). I'm giving them the benefit of the doubt. I'm giving them increased respect and I am getting it back (in return). I'm listening more. I'm listening to de-escalate potentially explosive situations, rather than contributing to their escalation. I'm able to approach a situation more calmly and talk through it. My co-workers and family have noticed changes in my attitude. I'm happier and a lot more easy-going. Relationships in my life have improved.

"The greatest tool for me was the RESPIRE 1 breathing CD. I use it 2-3 times a day to calm down and ease the situation. Since I've been doing this, my outlook and attitude have completely changed for the better. I can control my behavior and my anger. I didn't know I had the power to do this."

Indeed, many factors contributed to these positive changes over the course of 5 months and Steve experienced many trials and tribulations. Dietary changes, improved sleep, behavioral modification, EEG biofeedback, the support of his family, and a new way of breathing, all contributed to his success. He is sharing his story and teaching Coherent Breathing skills to his co-workers and others at the detention center.

4. Dodie *(family intervention)* is married and has 2 daughters and a 12 year-old son, Ryan. She had seen a Dr. Phil program on neurotherapy and alternative treatment of ADD/ADHD. Her search for a practitioner for her son brought her to the center. Ryan has significant learning and behavioral challenges, his behavior often disrupting both school and family environments. Both parents work, and the older daughter is often required to care for Ryan when mom and dad are working. In this case, the whole family was under significant pressure and needed intervention.

Ryan's NPA indicated elevated delta/theta and elevated high beta over frontal and motor cortex areas. Ryan's initial HRV was 25-30 beats with moderate but inconsistent coherence. After discussing the results with Dodie and her husband, a plan of action was developed including Coherent Breathing, HRV training, and EEG biofeedback, as well as behavioral and dietary modifications.

Dodie wanted Ryan to get started as soon as possible. It was fall break for Ryan and this provided the opportunity to begin the following morning. I requested that all family members attend the first session in order to introduce clearly-needed coping skills. (It is my finding that having all family members participate in the client's program dramatically increases the potential for a successful outcome.) In addition to some very basic behavioral and dietary recommendations, my intention was to educate them as a group on the importance of breathing and HRV in the management of stress.

The following morning, all family members did accompany Ryan to his first session. I provided a brief explanation of heart rate variability and how it is a key indicator of stress level. I further explained the role of breathing and its relationship to heart rate variability, and how breathing could be used to mitigate stress and anxiety. This explanation was provided in simple terms that everyone could understand, including Ryan and his 10 year-old sister.

I began with a discussion of Coherent Breathing. I played "Vocal Instructive Sequence" aloud and asked everyone to breathe together in synchrony with the rhythm on the CD. As everyone breathed, I assessed their comprehension and provided coaching. During this introduction to Coherent Breathing, Dodie's 19 year-old daughter volunteered to demonstrate the HRV program. In less than 1 minute, while breathing in synchrony with the CD, her HRV cycle demonstrated an amplitude of 30 beats with 79% coherence. She sustained this for 6 minutes. Immediately afterward, she reported "a feeling of calm in her upper body".

After this initial introduction to Coherent Breathing, each member of the family was given an opportunity to practice HRV training for 8-10 minutes, allowing them a brief time to consider the relationship between their HRV pattern and respective emotional states, as well as the connection with breathing.

During her 8-minute session using the "2 Bells" track with her eyes closed, the 19 year-old's HRV displayed an amplitude of 30 beats with 89% coherence. This time she reported "feeling relaxed and calm all over". She had never used HRV training previously, nor had she ever been involved in a conscious breathing practice. Other family members met with lesser degrees of HRV amplitude and coherence. All agreed to work with Coherent Breathing at home and use the HRV training program at the center when accompanying Ryan for a session. They left with two copies of RESPIRE 1, one for Ryan and one for other family members.

Days later, while at work, Dodie received a much-dreaded phone call from her daughter. Ryan was having a great deal of difficulty with one of his outbursts and was not able to go to school. This had happened a number of times previously and it was often the case that Dodie had to stop her work and return home. This day, Dodie asked to speak with Ryan. She said, "Ryan, remember the sound of the ticking clock? (Dodie was referring to "Clock & Bell".) Right now I want you to think of the sound of the ticking clock and let's breathe together." With that, she guided him to inhale and exhale for a few minutes. This short intervention allowed Dodie to successfully redirect Ryan to a state of calm and cooperation. Dodie completed her workday without further incident.

On Thursday, when Dodie brought Ryan in for his third session, she said that she was distressed and feeling overwhelmed. I provided her with a "personal" copy of RESPIRE 1 for use at work, and reinforced her understanding of how to use it. The following Tuesday, Dodie was calmer and more optimistic. Ryan's outbursts had decreased from 1 or more per day to zero in the prior 8 days. These are Dodie's thoughts in her own words:

"The effects of breathing for me have been rapid… calming. This practice has allowed me to focus on my life and not the agony of each day. I started the breathing technique about 2 weeks ago. The most profound thing I have noticed is the calmness – I have been able to maintain a sense of calm on a daily basis. I have been able to diffuse my own anxieties as well as help my children through theirs. This is not something that has come lightly or easily for me in the past. I've also found that situations at work that normally would have been major crises are now merely events in my day. Co-workers have even noticed a change in my attitude and how much more pleasant I am in general. I'm happy I came here to find Coherent Breathing and thank God for this peace of mind."

Dodie and other family members were encouraged to participate in Coherent Breathing and HRV training whenever they accompanied Ryan to his sessions. Dodie does HRV training for 8-12 minutes twice a week. On October 13th she commented, "I felt tired and drained when I started. When I finished I felt calm and my mind clear" – her HRV amplitude was 15 beats with 84% coherence. Her comment of November 10th, "Awesome" – her HRV amplitude was 15 beats with 100% coherence. Dodie reports practicing Coherent Breathing daily. She has become so proficient that she is now teaching her 10 year-old daughter.

Ryan continues with EEG biofeedback. Coherent Breathing and HRV training are an integral part his treatment program.

The New Science of Breath

Beginning his session with HRV training provides me with an opportunity to assess his autonomic status, and provides him with the opportunity to unwind from his school day prior to EEG neurotherapy. As previously noted, Ryan's initial HRV demonstrated an amplitude of 25-30 beats with 42% coherence. Two weeks later, while using RESPIRE 1, Ryan's amplitude was 30 beats with 51% coherence. His comment at the end of this session was "Great!" One month later, 6 weeks into his treatment program, Ryan's HRV was 30 beats with 85% coherence. At the end of this session he commented, "Feeling very calm". While Ryan continues to request the use of RESPIRE 1 during HRV training, the focus is on cultivation of positive thoughts and emotions. He is demonstrating progress.

Over the course of 8 weeks, Ryan's EEG showed a significant decrease in previously elevated frontal delta/theta and high beta frequencies. With encouragement, Ryan continues to practice Coherent Breathing at home and at school, and has personally elected to incorporate it into his speech therapy sessions. He is more open to intervention and redirection, and his relationship with family members is improving. We are all hopeful that with continued progress and therapeutic support, he will reach his personal and academic potential.

5. Janet *(traumatic brain injury)* is a bright, energetic 55 year-old who had suffered traumatic brain injury in an automobile accident. She had worked as a flight attendant for a number of years prior to her injury and, while she has not been able to return to her prior career, she is living a very full and productive life.

Janet has been a client for many years; she and I know each other well. With all of her neurocognitive training completed, we had settled into a maintenance routine of perhaps 8-10 EEG biofeedback sessions every 3-6 months, if needed, for what she called "touch ups at the magic shop". Over this course of time, we had established a pretty solid understanding of the EEG norm where she functioned best.

At this time, Janet was in town planning the sale of her Dallas home. It had been a year since her last visit to the center. Prior to any intervention, I asked Janet for an update. What was giving her the most difficulty? I noticed that she was more anxious than at any other time that I had seen her. The stress of building a new home and relocating to the west coast resulted in a recurrence of prior symptoms including decreased word finding, attention difficulties, forgetfulness, and altered sleep despite sleep medication. I did an HRV and EEG "update". Her HRV amplitude approximated to 7 beats and her coherence ranged from 8-46%. Her EEG demonstrated elevated delta /theta (2-7 Hz.) as well as elevated high beta in frontal as well as left and right hemispheres along the motor cortex. Her breathing was relatively rapid and shallow.

Janet wanted to move quickly and stated she needed to "function with mental clarity" in order to accomplish all tasks that were before her. Having observed her HRV and breathing pattern, I proposed that we begin with Coherent Breathing. With coaching, she acquired the skills rapidly and was able to demonstrate the method effectively. We elected to use Coherent Breathing in conjunction with her first EEG session during which "Vocal Instructive Sequence", followed by "Clock & Bell", played softly through headphones. Following her session, I provided her with a copy of RESPIRE 1 for use at home. The next day, Janet reported feeling a bit clearer and having more energy. She said that she had practiced the breathing method overnight and "it seemed to make a difference".

One of the concerns for Janet is that, when she felt good, she often overdid it and her symptoms reoccurred. By day 3, she had overextended herself with some new-found energy, and was racing around at 90 miles an hour. On this third day, her breathing had returned to a short, shallow pattern and she felt exhausted. We started her session with Coherent Breathing and I observed her – she was hyperventilating, lifting her shoulders with each inhalation, forcing her breathing. She noted tingling in her fingers and toes and felt light headed. We paused for a few minutes and began again. This time I reinforced her breathing tempo by raising and lowering my hand as if to draw a circle in the air. In a short

The New Science of Breath

time, she was calmer and ready to start the EEG biofeedback session.

By session 4, Janet was doing RESPIRE 1 at the start of her day while she organized the daily activities on a list. She became very deliberate about pacing herself and setting limits for her daily activities. During this session, her HRV amplitude was in the range of 18 beats with coherence ranging between 50 and 98%. Her EEG had also changed dramatically from that of the first session.

Once Janet had experienced and understood the effects of Coherent Breathing, she at times asked to use it throughout her session (with eyes open). It was during one of these sessions that I recognized what I now know to be a second pattern resulting from Coherent Breathing. Within 15 minutes, Janet's high beta and delta/theta had decreased by 1-2 microvolts and the EEG graph was more organized. (This pattern was very consistent with what I had seen during Steve's EEG biofeedback session.) At the same time, Janet reported feeling "the shift". (I have a number of clients who state during sessions that they can feel the exact moment when they shift into the state of clarity and increased focus. Janet is one of them.)

Over the course of 14 sessions occurring over 8 weeks, Janet's EEG pattern changed, resulting in improved mental clarity and sleep. Her HRV also changed significantly. On September 27th, her first visit, her HRV amplitude was 7-8 beats with coherence ranging between 8 and 46%. Two months later her HRV amplitude was 15-18 beats with a much steadier 97% coherence.

With an increased ability to express her thoughts, Janet commented that the "pacing" of "Clock & Bell" enabled her to be "consistent" with her breathing. Janet commented that she has not felt this calm and relaxed in well over a year and attributes this calmness to her new-found breathing skill.

For continued success and sustained benefit Janet continues to practice Coherent Breathing and will resume use of her HRV program when she returns to the west coast. To reinforce this she has written reminders in her day planner.

She will return home with a new life skill – Coherent Breathing.

* * *

Several of the clients with whom I'm working have addiction challenges, and I elected to use RESPIRE 1 prior to, and at times during, their EEG sessions. While working with them, a third pattern resulting from Coherent Breathing became evident. When breathing in synchrony with "Clock & Bell" or "2 Bells" during eyes-closed training, clients rapidly access the alpha state, a process that can often take much longer to achieve. The next two observations demonstrate this.

6. Jake *(attention disorder & addiction)* is 22 years old. He has diabetes and has been dependent on insulin since childhood. He came to the center after being discharged from the hospital where he was in a coma after his blood sugar had spiked to 700. Jake had an addiction to crystal methamphetamine, the use of which had a profoundly negative effect on his diabetes. Obviously, Jake's parents were concerned, and at the encouragement of his counselor they came in for an evaluation to assess the potential benefit of biofeedback therapy. Jake was scheduled to enroll in a residential addiction treatment center in the coming weeks, but he found it difficult to focus and concentrate, and was concerned that this would impede his progress while in treatment. His therapy program was to be short and intense.

Jake's NPA demonstrated elevated frontal delta/theta as well as elevated high beta in occipital areas. Jake's blood sugar levels were not yet stable. He was clearly restless. Given his report of difficulty with concentration, a TOVA was also done and the results were suggestive of both hyperactivity and an attention disorder. Based on Jake's evaluation and goals, we started his program with Coherent Breathing and HRV training augmented with autogenics and relaxation skills. Jake also agreed to check his blood sugar level at the start of each session.

Initially, Jake's HRV amplitude was 7-8 beats and was completely erratic. He was unable to achieve even the slightest degree of coherence. He was then trained in the practice of Coherent Breathing and I had him perform it without the HRV program, as his early inability to make much progress added to his frustration. I provided a copy of RESPIRE 1 and asked him to practice at home. Over the course of several days his comprehension of the method grew and he was able to employ it in the moment. When his command of Coherent Breathing was clear to me and he could perform it without coaching, we resumed the use of the HRV program, this time while Jake breathed in synchrony with RESPIRE 1. Gradually Jake developed a command of his HRV rhythm, his amplitude ranging between 11 and 20 beats with 74% coherence. With this progress, he was able to begin alpha/theta EEG protocol for addiction and achieved some success prior to his departure.

Jake was away 6 weeks. Within a week after returning he resumed his neurotherapy sessions. I noted that he was restless and upon assessing his HRV found that his amplitude was 7 beats with very low coherence, not unlike his HRV 9-10 weeks earlier. I reinforced the importance Coherent Breathing and he was quickly able to reassume his previous level of competency. With this, the basic session protocol of Coherent Breathing, followed by HRV training, followed by alpha/theta EEG biofeedback, was resumed.

Initially during alpha/theta training, Jake would fall asleep and had difficulty making significant progress. I elected to use RESPIRE 1 to keep him awake. Here again, I was able to see effects of Coherent Breathing that I had seen before. Each time we used this method, Jake was able to shift immediately into the alpha state followed shortly thereafter by the desired alpha/theta state. The high frequency beta that had been prominent at the occipital lobe just prior to his use of Coherent Breathing with eyes open and closed had decreased. With this advance, his alpha/theta training became much more productive and we were able to achieve desired goals quickly. For the last few appointments, I've incorporated "2 Bells" into Jake's training sessions. He immediately shifts into the alpha calm state. He no longer falls asleep during the sessions and is able to attain the alpha/theta state with vivid imagery and good recall.

Jake attended sessions once or twice per week over a period of 10 weeks. During this time, his HRV has demonstrated dramatic improvement, his amplitude increasing from 7-8 beats with indiscernible coherence to its present 28-30 beats with high coherence. He is presently employed, and working with an addiction counselor on an ongoing basis, and is aware that managing his addiction is a long-term process. Jake reports that he uses RESPIRE 1 to reinforce Coherent Breathing "almost daily" and, at times, to help him go to sleep. He will continue to attend sessions once per week until the end of this year. At the end of Jake's most recent session, while reviewing his progress he volunteered the following:

"The (drug) craving has decreased. I'm calmer. I've been able to control my temper. My attitude and mood have improved. My blood sugar is stable and I have been able to decrease my insulin doses." Jake wanted to share his story and noted, "If my experience can help someone else, that would be great."

7. Erin *(attention disorder & addiction)* is a 21 year-old young lady who is bright and energetic on the surface but reported being very shy and having very few friends. She came to the center in late August as the result of an addiction to hydrocodone and use of crystal methamphetamine for a period of 6 months. The family was particularly concerned because it had led her into some recent legal difficulties. Erin's mother accompanied her to the initial assessment as well as the first day of treatment.

The complaints that Erin presented were of anxiety and internal discomfort. She had difficulty going to sleep, which could take her 60-90 minutes. She had difficulty with reading and recall, her mind was often wandering. This, she reports, had been present most of her life. She would sit in the back of the room worrying what people would think about her, and often heard music in her head. Her attempts at college immediately after high school did not meet with success. She returned home and continued the use of hydrocodone and crystal meth.

An NPA and TOVA were completed and an extensive history was taken. The results of her testing were suggestive of an attention disorder with hyperactivity, and there was some evidence of depression and anxiety. A treatment plan was developed calling for Erin to participate in 3 sessions per week.

The initial objective was to address her feelings of anxiety and internal discomfort and to improve her sleep pattern as well as her ability to focus. While we outlined the need to address the depression and addiction, she felt it was more important for her to be able to sit in the classroom and pay attention.

Erin's treatment program began with instruction in Coherent Breathing. At first, she found breathing in this way to be difficult and laborious, but as she persevered she began to develop a sense of calm, this becoming a consistent characteristic of Coherent Breathing. Once she was reasonably proficient, we progressed to HRV training. Here again, she found the practice to be tedious. Three minutes into her sessions she would report feeling bored, frustrated, and sometimes angry. Her initial HRV amplitude varied between 7 and 20 beats with inconsistent coherence. Within days, she gradually became more accepting and less resistant to the practice. During this period, her HRV amplitude and coherence gradually increased.

She began EEG biofeedback with an initial focus on improving sleep, later shifting focus to improving attention. Each session started with autogenics and relaxation skills in combination with Coherent Breathing and HRV training in order to reduce her anxiety. Within 4 sessions, her ability to use the HRV program for extended periods increased and she no longer reported being "bored". By this time, Erin's HRV was approximately 30 beats with 91% coherence. She was growing calmer and more comfortable, along with gradual improvements in her breathing and sleeping patterns.

With this progress, we began the alpha/theta protocol for addiction. It was a very positive experience for her. She chose to use her favorite RESPIRE 1 track, "2 Bells", during the session. Observing her EEG response, I noted that she, like Jake, immediately shifted into a quiet alpha state followed closely by the desired alpha/theta state. Again, this state was accompanied by a pronounced decrease in high beta in the occipital area. During this and following sessions she achieved both deep relaxation and vivid imagery.

From late August, Erin attended every one of her sessions and with each session she became brighter and more engaged.

Her mental outlook has improved. She has increased eye contact, is more assertive, and has increased interaction with her peers. She is attending classes at a local college with the goal of increasing her grade point average and eventually attending a larger university. She reports no further drug use.

I asked Erin, relative to her therapy experience, what in her view had influenced her most significantly. She responded immediately, "It's the breathing". Coherent Breathing, HRV training, and EEG biofeedback have all contributed to Erin's improvement. However, the most important ingredient in her success was her conscious effort to apply what she learned during her sessions to her daily life – specifically, Coherent Breathing. Today Erin's HRV amplitude is 28-30 beats with 100% coherence.

8. Ashley *(stress & anxiety)* is a bubbly 10 year-old 5th grader and only child. She was referred by a former client who thought Ashley might benefit from EEG biofeedback for what she referred to as "test anxiety". While Ashley's grades were good, tests were unusually challenging for her. In Ashley's words, whenever she took a test, her brain would just "shut down". Ashley had also had nightmares that prompted her to want to sleep in her parent's room. Her parent's wanted Ashley to overcome her fears and to be able to sleep in her own room.

During the intake interview, it was noted that Ashley had some learning difficulties, specifically relating to the retention of information in the classroom environment. When I questioned her, it was difficult to understand her. Her eye contact was poor, and she didn't open her mouth very much when she spoke.

Ashley's initial EEG assessment revealed elevated high beta as well as elevated delta/theta in frontal areas. Her electro-dermal response was extremely elevated during cognitive tasks, and she found it very difficult to sit still. After reviewing her assessment results, I met with Ashley's parents, and together we came up with the game plan which included helping Ashley to feel more confident and safe in her environment. Ashley would attend sessions twice per week and these would involve multiple treatment methods.

The New Science of Breath

The program began with a focus on lowering Ashley's electro-dermal response (EDR), which was extremely high, in the range of 20 micromhos. I elected to employ Coherent Breathing for this purpose and we did several consecutive sessions simply focusing on breathing. Within the course of 3 sessions, Ashley's EDR "score" dropped into the 10 range. We then shifted our focus to HRV training. Ashley's initial HRV amplitude varied widely between 10 and 28 beats with coherence ranging between 31 and 74 percent. Initially, HRV training proved challenging for her but she was quick to grasp the relationship between her breathing pattern and her HRV waveform, and with this understanding her consistency improved quickly.

With significant EDR and HRV enhancements, we shifted the primary focus to EEG biofeedback, specifically the reduction of high beta and delta/theta in frontal areas and along the motor cortex. Each session continued to begin with the practice of Coherent Breathing followed by HRV training. With this basic session protocol, within 6-8 sessions we started to see desired changes in her EEG. With these changes, Ashley began to report an increased ability to focus and retain information in the classroom. She was leaving sessions brighter and more confident, smiling and saying "Thank you". And, she is sleeping alone in her own room (with the exception of the occasional visit from her canine buddies).

As of the end of November, Ashley's HRV amplitude is 28-30 beats with 100% coherence. She reports being calm and focused in the classroom with better recall. She has better eye contact, is speaking more clearly, and is more assertive. She has learned to use breathing "in the moment" to feel less anxious when she is confronted with stress, including prior to testing. Recently, in the context of discussing her progress with Ashley and her mother, I asked, "Ashley, what do you think has helped you most?" Her answer was, "It's that breathing".

This past week Ashley gave me a copy of her progress report for the past 6 weeks. She and her parents were beaming. She has demonstrated improvement in all but one subject, with all subjects averaging 90 or above.

The frequency of Ashley's sessions was decreased to once per week during October and her therapy program will conclude in December. Her parents plan to purchase an HRV training program for home use, as well to continue the application of Coherent Breathing.

9. Ann *(anxiety & "panic attack")*. Late one night I received a call from Ann, the mother of a client that I had worked with previously. Ann, 48, is devoted to her children and spends much of her time making sure their special educational needs are met. Ann had just seen her internist. She reported that 2 nights ago she had "a horrible feeling in her chest" and felt like she was "going to die". She said that her heart had been racing and she was worried that she might be having a heart attack. This had happened more than once. The internist ran several tests and reported that her EKG and her 24-hour Holter-monitor (EKG) were normal.

As Ann remained concerned, her internist suggested that she also undergo a cardiac stress test. Again, the results were normal and her doctor advised her "not to worry about it". But, Ann remained extremely frustrated and worried. Her son had undergone ablation for heart arrhythmia and she was worried now that she may also have a cardiac problem.

Still concerned and in need of answers, Ann asked if I would meet with her and go over her diagnostic results. I said that I would be pleased to, but that I would defer any recommendations regarding her treatment to her physician. She had been reading diligently looking for answers, and showed up at the center with her arms full of books. We began by going over what was going on in her life. She related that she had been very busy and under a great deal of stress, that many things were going on in her personal life and the lives of her children. We reviewed her lab results, as well as medications she was using and nutritional supplements she was taking.

Using heart rate variability as a tool to assess her autonomic status, I monitored her HRV for a few minutes. She was extremely restless and her HRV cycle was totally erratic with very low coherence. She and I agreed that it would be most beneficial if she could calm down so she could consider her situation objectively and determine what she wanted to do next. I had provided a copy of RESPIRE 1 for her son's use in February, and asked her if she had tried using it. She had not.

We sat down and I instructed her in the practice of Coherent Breathing. As we sat and worked with her understanding of the relationship between her heart rate and her breathing she suddenly stopped and said, "You know, I think I stopped breathing during those episodes. I felt my tongue go behind my teeth and I just stopped breathing!" The next words out of her mouth were, "Was I having a panic attack?" I responded that I could not be sure but that it was possible.

We resumed instruction in Coherent Breathing while monitoring her HRV rhythm, and over the course of minutes, as her comprehension of the method improved, her HRV amplitude increased from 4-5 beats at 18% coherence to 9-10 beats at 28% coherence. One week later Ann requested a follow-up visit. I checked to see what her HRV looked like without the use of the CD. Both her amplitude and coherence had diminished as compared to a week earlier. When I suggested that she elicit a positive thought or emotion, she said that "it was difficult for her". Again, we employed RESPIRE 1. Shortly thereafter, she began to experience the sense of calm and relaxation. Needless to say, she found this very comforting. She commented, "I felt stressed at first. I could feel it when my heartbeat skipped. After starting the CD I felt more relaxed and my heart rate evened out."

Ann came in for another session a week later. This time she had been practicing her breathing. She got into the zone within 1 minute and stayed there without the use of the CD. Her HRV demonstrated an amplitude of 10-12 beats with 86% coherence!

After this experience, she decided that what she needed to do was carry a copy of RESPIRE 1 with her and use it whenever she felt an "attack" was imminent. She left with the intention to load it onto an MP3 player so she would be able to use it at any time and any place. I encouraged her to return to her internist and to get answers to her questions. I also requested that she keep me updated.

Over the next couple of days, she did call to report that she was using RESPIRE 1 and it was helping her function better. She noted that she was feeling calmer and more confident, and that some of the stresses in her life had abated.

Ann's rapid response to Coherent Breathing is what I am beginning to call "RESPIRE to the rescue". While Coherent Breathing did not resolve Ann's concerns about her health, it did allow her to regain her composure and peace of mind so she could see things more objectively. During a recent phone call with Ann she said, "Doctors really ought to play this CD in their office. It would help their patients stay calm during times of stress."

Ann has purchased an HRV training program for home use and continues to practice Coherent Breathing.

10. Andrew *(attention disorder, depression, & anxiety)* is 17 years old. He came to the center in mid-August to be evaluated for EEG biofeedback. He was accompanied by his mother, who expressed concern over Andrew's poor grades, inability to focus, decreased socialization, constant complaint of fatigue, and difficulty getting up in the morning in order to make it to school on time. During his assessment Andrew appeared quiet, had decreased eye contact, and spoke very little during the first part of the interview. Andrew is the oldest of 4 children and has a sibling who is severely disabled. Andrew experienced difficulties in school at an early age and had been on medication for attention deficit disorder (ADD) since 7th grade. He has asthma and was taking a sustained release asthma medication. He was also using an acne medication which he had recently changed due to a severe allergic reaction, a symptom of which was depression.

The results of Andrew's NPA were indicative of anxiety, depression, and ADD. His EEG exhibited elevated high frequency beta in the 19-28 Hz. range, both frontal and occipital, as well as elevated delta/theta in frontal areas. HRV assessment was deferred until his first session. I outlined the course of therapy and we discussed other factors including nutrition, social skills, sleep pattern, etc. We agreed that Andrew would

The New Science of Breath

attend 3 sessions per week for the next month, after which we would re-evaluate his progress.

Andrew's HRV was assessed during his first session, and his average heartbeat rate was 108 beats per minute but was also quite variable. His HRV amplitude averaged 5-15 beats and was extremely erratic. His initial coherence was 32%. He mentioned that he could feel his heartbeat and that it was irregular toward the end of each day when the effect of his ADD medication was tapering off.

I began to instruct Andrew in the practice of Coherent Breathing, provided a copy of RESPIRE 1 for his home use, and encouraged him to practice the technique regularly. Sessions 2 and 3 began with Coherent Breathing and HRV training. In session 3 we focused exclusively on Coherent Breathing and HRV, employing "Vocal Instructive Sequence". We also practiced body scanning techniques, consciously relaxing face, hands, feet, etc. In this 30-minute period, Andrew's heart rate was 92 beats per minute and his HRV amplitude was 25 beats with 95% coherence! At the end of 20 minutes, he looked up and said, "I feel weird – I'm very calm." Andrew had never experienced this feeling before.

The initial EEG training was focused on reducing the elevated frontal delta/theta and high frequency beta components, followed by depression protocol and management of his anxiety. Within a few sessions Andrew began to report that he was feeling more alert and that his ability to focus was improving. A total of 8 EEG sessions were completed, after which Andrew and I reassessed his progress and discussed the importance of beginning his training for depression. Andrew requested that one of his 3 sessions per week be devoted to focus and concentration. At the time of this writing Andrew has completed 6 sessions for depression and 2 additional sessions for attention.

His parents and I speak frequently and we have all noted a greater sense of confidence, improved eye contact, and enhanced ability for Andrew to express himself. His parents reported that he was engaging more proactively in school, and that his friends and family have noted a positive change in his behavior. Andrew's dad recently called to say that during the

2nd six-week grading period Andrew had passed all of his classes and had a 100 in algebra!

At our last session, after Andrew completed his HRV training, he related that he had seen his pulmonologist recently and that his doctor noted that his asthma as well as his overall pulmonary function had improved. I congratulated him and upon discussing his doctor's comments Andrew stated that he is confident that this improvement is due to his regular use of Coherent Breathing and HRV training.

Over the course of 3 months, Andrew's sessions have decreased to 2 days a week. His grades continue to improve and both his depression and anxiety have decreased. His present HRV amplitude is consistently 25+ beats with 95% coherence.

11. Peter *(depression)* is a 50 year-old father of four and independent businessman. In the last 2 years, Peter has received radiation and surgery for non-Hodgkin T-cell lymphoma (a cancer of the immune system) which is now in remission. He has a past history of depression but used antidepressant medications for only 30 days before deciding that "they really weren't for him". Peter's youngest daughter has been severely disabled since birth with a seizure disorder that is now under control. His 17 year-old son is receiving EEG biofeedback here at the center. Peter describes himself as "a constant worrier".

Peter had accompanied his son to a number of sessions and, as is the case for all parents, I encouraged him to participate in heart rate variability training and Coherent Breathing. I instructed Peter in both techniques and he would practice while his son was in session, with me checking in on him to provide coaching.

Peter's initial HRV amplitude averaged 8-10 beats with 22% coherence. He found HRV training to be a struggle, seemingly unable to establish a sense of cause and effect with his HRV cycle. He related that he was under a great deal of stress and that his depression seemed to be returning. He wondered if he might benefit from neurotherapy. I suggested that, while I would be glad to work with him, he consult his MD. Two days later, at the encouragement of his wife, Peter returned for an EEG neurofeedback evaluation. He noted that he had seen improvement in his son and wanted to know if EEG neurofeedback was a treatment option for him.

Peter's NPA demonstrated high EDR in the range of 19-26 micromhos, as well as elevated high beta in the 19-28 Hz. range in frontal and occipital areas (his biometrics are very consistent with his son's). He revealed that his sister had suffered anxiety and depression for most of her life and died as the result of alcoholism and street drugs, this knowledge adding to his concern that he and his family may share this pattern. A treatment plan was established – Peter would participate in 3 sessions per week with the initial focus on reducing anxiety. I also encouraged him to follow up with his MD and the next time he had lab work done to have his thyroid and testosterone levels assessed.

Peter elected to resume use of his antidepressant. He related that his stress was building, due to his impending quarterly cancer checkup as well as ongoing concerns for his family and business. I provided Peter with a copy of the RESPIRE I CD for use at home and in the office. With this, he began practicing daily. The next time I saw him, while his comprehension of Coherent Breathing had increased, he still had not integrated it into his regular breathing pattern and was not deriving the potential benefit. For this reason, we made HRV and breathing the focus of his second session. By the end of this session he commented, "I can do this!" referring to his having discerned the "connection" between his HRV and breathing cycle. I encouraged Peter to employ Coherent Breathing during his normal daily activities and particularly when he was feeling anxious or depressed.

In his third session, attempts at "eyes closed" autogenic training met with some success and he was able to reduce his EDR slightly, but concerns about his impending cancer check-up and difficulties at home continued to present themselves during his sessions. He noted that, "while my heart is in this, my mind won't cooperate".

In Peter's fourth session, we continued to focus on Coherent Breathing. While breathing in synchrony with RESPIRE 1, Peter achieved a new level of relaxation, decreasing his EDR from 18 to 11 and increasing his hand temperature to 90 degrees. With this breakthrough, he was receptive to attempting the eyes-closed alpha/theta training for anxiety. Peter was OK with this, but found the idea of using a suggestive voice script to facilitate relaxation unappealing. As an alternative, I elected to use "Clock and Bell" during his session. Before starting the CD, I assessed his brainwave pattern, first with eyes open and then with eyes closed. Within 5 minutes of turning on the CD and Peter beginning to breathe in synchrony with it, his alpha amplitude increased significantly. A few minutes later his EEG pattern clearly reflected that he had settled into the alpha state, followed shortly thereafter by the alpha/theta state. At the end of the session Peter reported he had experienced vivid visual imagery and felt "amazingly calm".

As of this writing, Peter has had 4 alpha/theta sessions employing the use of Respire 1, and during each session he was able to elicit the desired EEG and physiological response. He reports an increasing feeling of "calm" and "a sense of peace". His high beta amplitude is gradually decreasing, and with it he reports feeling less stressed and more focused. His test results are back and the cancer remains in remission. Peter and his family are now in family therapy and he reports that all family members are communicating more effectively. He is able to face his fears and concerns with less stress and "fewer anxious moments" and feels "more in control" of his life.

Peter continues with his neurotherapy. He has created a quiet place in his home were he can retreat and practice Coherent Breathing daily. He has developed a firm grasp of the relationship between his HRV and breathing patterns, and is learning to correlate the state of his heart rate variability cycle with his internal feelings, enabling him to manage his stress more proactively. In late August, Peter's HRV amplitude was 8-10 beats with 22% coherence. In early November his amplitude was 10-12 beats with 91% coherence. Over this same period his EDR declined by 10-15 points.

12. Jacob *(attention disorder & depression)* is 10 years old and in the 5th grade. He was returning to the center to resume neurotherapy for attention and behavioral difficulties after having taken the summer off. He had previously been diagnosed with an auditory processing disorder. Jacob's parents expressed concern about the difficulties he was now having both at school and at home, including not completing his work, failing to turn in completed assignments, and not doing his assigned chores at home. Jacob expressed frustration with socializing and peer relationships, and was often upset and at times angry. His parents noted that he was easily moved to tears and when this happened it was difficult or impossible to console him.

I had not seen Jacob for several months, and I did a neurophysiologic update including an EEG assessment for depression, which was affirmed. A TOVA and an HRV assessment were also completed. Jacob's HRV demonstrated inconsistent amplitude which approximated to 10 beats with low coherence. During his assessment, Jacob noted that he was unable to think of even one positive thought. I reviewed the information with Jacob's parents and a course of treatment was outlined, the focus of which would be EEG alpha asymmetry protocol for depression.

Jacob's sessions were to begin with Coherent Breathing and HRV training. Initially, this did not go well. Jacob was unable to effect any change in his HRV. Again, when I encouraged him to think of a happy thought or recall a happy moment in his life, he replied, "I can't find anything to be happy about". I attempted to teach Jacob Coherent Breathing, and while he would listen to the CD he made no attempt to synchronize his breathing rhythm and he was apathetic about trying.

Prior to Jacob's second session, he noticed two other 5th graders, Samantha and Ashley, who had just completed their HRV training sessions, both with high amplitude and coherence. He was impressed with Ashley's session which demonstrated 30 beats with 100% coherence! I could hear Jacob and Ashley talking. She and Samantha were coaching and encouraging Jacob and instructing him how, "by following the man's voice on the CD" (in reference to "Vocal Instructive Sequence") he could also "have a high score". When I inquired, Jacob expressed interest in trying it again. I allowed him to select which track he wanted to listen to and he chose "Vocal Instructive Sequence". Jacob wanted to use the headphones and, after adjusting the volume, the only instruction I gave him was to "follow the man's voice and do whatever he tells you to do".

In the course of 5 minutes, Jacob's HRV demonstrated a beautiful sinewave with an amplitude of 25 beats or higher with 68% coherence. (Jacob's HRV approximated that of Figure 8.) He continued this for 24 minutes with his eyes open. The only interruption was his request that his father come into the room to see his progress. His dad was brought into the room and was amazed at Jacob's ability to maintain the high level of amplitude and coherence. Jacob's comment at the end of that HRV session was, "The best – It's a comeback – Wahoo!" Jacob shifted easily into his EEG session for depression and rapidly acquired the desired state, ending his session feeling brighter and happier.

Jacob's dad had previously participated in HRV training while at the center, and he was so impressed that I offered him the opportunity to learn Coherent Breathing and give it a try. He agreed, and after taking a few minutes for me to explain and demonstrate the method, he started the program. This time, while listening to RESPIRE 1, within 2 minutes his HRV was more robust and coherent than he had achieved at any time previously. Jacob and his dad took home a copy of RESPIRE 1 so they could practice Coherent Breathing between sessions.

At the beginning of Jacob's next session, again doing HRV training while listening to "Vocal Instructive Sequence", his HRV amplitude was again 25 plus beats, this time with 86% coherence, indicating that his comprehension of Coherent Breathing was sound.

Jacob has completed 5 EEG biofeedback sessions for depression and is responding quickly. He is smiling, reports feeling better, and says "Thank you" at the end of each session. His parents and I are now working a broader plan to enhance his learning environment and improve his socializing skills.

The most profound observation is that, at the urging of his peers, Jacob established a strong connection between his breathing and HRV during a single session. With this, he was quickly able to access and maintain a relaxed state that he carried into his EEG training. Now he is able to achieve success with the HRV program without the aid of RESPIRE I, focusing on cultivating positive emotions.

Jacob's HRV has consistently improved, his present amplitude being 30 beats with 90-95% coherence. It is apparent that this improvement in his HRV rhythm and autonomic state has also enhanced the efficacy of his EEG biofeedback sessions.

* * *

To demonstrate the efficacy of Coherent Breathing, I elected to include clients with a wide range of ages and diagnoses including anxiety – general and performance, hypertension, pain, anger, depression, attention disorders, behavioral disorders, addiction, "panic attack", traumatic brain injury, and sleep dysfunction. In some cases these conditions also involved underlying health conditions including cancer and diabetes. Ages ranged from 10-55.

In my estimation, the results have been remarkable, specifically the accelerated rate of progress when employing Coherent Breathing. Most of them reported that they experienced dramatic improvement with significant lessening of symptoms within a few weeks of beginning the breathing practice. When the client embraced the practice of Coherent Breathing and especially when they adopted it as their primary mode of breathing, this process usually required no more than 2-3 sessions. In my neurotherapy experience, these timeframes are atypical, especially in the light of some of the conditions and their severity presented in these clients. It is my belief that this accelerated pace was a consequence of establishing optimal autonomic nervous system balance *prior* to neurotherapy.

In most cases, Coherent Breathing was used in conjunction with heart rate variability biofeedback as a first step to elicit and establish a viable state of autonomic nervous system balance prior to undertaking other forms of biofeedback. With autonomic nervous system balance having been established via breathing, I noted that all methods, including EEG biofeedback, were significantly more productive. "Accessing alpha" is a good example. When an EEG protocol involving alpha was indicated, the client was often successful in eliciting that state *immediately* – in less than 60 seconds after Coherent Breathing begins (once the method is understood). Again, in my experience, this is unusual, especially for clients that initially exhibit strong sympathetic dominance.

During this adjunctive use of Coherent Breathing, I noted the following distinct phenomena:

- The development of a profound sense of "calm" and "inner peace" within minutes of breathing coherently,
- A rapid and often impressive decrease in delta/theta components at relevant sites,
- A rapid and often impressive decrease in high frequency beta components at relevant sites,
- Accelerated access to the alpha state followed soon thereafter by the alpha/theta state (when employing relevant protocols),
- A rapid yet appropriate reduction in EDR, and often
- An equally rapid and dramatic increase in hand temperature.

It is apparent that these changes are readily facilitated during the state of autonomic balance.

In summary, I have used Coherent Breathing for 11 months in multiple settings with well over 100 clients and family members. At first, I did not fully appreciate its potential but now, having seen the efficacy with numerous clients regardless of age or condition, it is clear that there is something very fundamental at work here. As the number of clients in my "database of experience" continues to grow, it is becoming more evident that sub-optimal breathing is playing a major part in many of their conditions, regardless of age.

All of the individuals in the case observations have succeeded in changing the quality of their lives through conscious effort. Coherent Breathing was their primary tool. Where there was comprehension, consistency, and compliance of breath, results were evident:

The New Science of Breath

- Maria has <u>experienced peace</u>.
- Joseph found <u>comfort within himself</u> after searching for many years.
- Steve found <u>a power that he never knew he had</u>.
- Dodie discovered <u>calmness and peace of mind</u>.
- Ryan has learned <u>to modify his behavior</u> by breathing!
- Janet found <u>a new pace</u> that helps her govern her life.
- Jake has <u>a new ally</u> in his fight against addiction.
- Erin has <u>newfound confidence</u> in meeting her academic and life goals.
- Ann has an <u>anchor</u> to help her in times of crisis.
- Jacob found <u>a moment of greatness – a happy thought</u>.
- Andrew has developed <u>increased self assurance</u>.
- Peter understands <u>cooperation between his heart and mind</u>.
- Ashley discovered…..<u>it's that breathing!</u>

Clients, parents, friends, and families, whom I have trained in Coherent Breathing, have taken away more than a quick fix – they have taken away a fundamental life skill. Learning how to breathe coherently is like learning to ride a bicycle. Once you learn the skill, you have it for life.

If, in my comparatively small neurotherapy practice, clients who I have worked with have experienced such positive benefits, my thoughts would be:

- Would Coherent Breathing not be of great value to my friends and colleagues, not only in neurotherapy but in other fields such as psychiatry, cardiology, neurology, obstetrics, pain management, home health, etc.?
- Where is autonomic nervous system balance not of value?

The journey begins with breath…

7

Yoga & Meditation

"The yogi's life is not measured by the number of his days, but by the number of his breaths." B.K.S. Iyengar – **Light on Yoga**[13]

The moment of autonomic nervous system balance and cardiopulmonary resonance is subtle. One can experience it in passing as a fleeting moment of intuition, peace, comfort, or oneness. It isn't until you sustain the moment for a matter of seconds or minutes that you begin to understand its potential. And for this reason, it has remained an esoteric mystery until now. It is when it is incorporated into your daily life as your normal breathing pattern, that it is revolutionary, or should I say evolutionary. The very practice brings about change, very much as the practice of yoga or meditation brings about change, for I argue that autonomic nervous system balance is the master key to yogic attainment. Therefore, when Coherent Breathing is incorporated into your daily life, you are in effect "doing yoga" all the time.

One of the most intriguing moments in the investigation of this "new science" was the point of realization that it isn't really new – it's just that the territory isn't well charted.

The age-old science of yoga leads one toward autonomic balance by both promoting balance and countering the effects of imbalance. This includes the primary yogic methods of breathing (pranayama), physical exercise (asana), and meditation (dhyana).

The New Science of Breath

Pranayama is the cornerstone of yogic science. Ultimately, it is the foundation of all yogic attainment, for without cultivating the breath nothing else is possible. As such, an explicit objective of pranayama is autonomic nervous system balance. In *Light On Yoga*, B.K.S. Iyengar says, "Emotional excitement affects the rate of breathing; equally, deliberate regulation of breathing checks emotional excitement. As the very object of yoga is to control and still the mind, the yogi first learns pranayama to master the breath. This will enable him to control the senses and so reach the stage of pratyahara. Only then will the mind be ready for concentration."[14]

A primary function of hatha yoga, characterized by the practice of various asanas (physical postures), is to bring the practicer face to face with the challenge of autonomic nervous system balance, cultivating conscious control over both sympathetic and parasympathetic response. We often speak of the body as being "tight" or "inflexible", when in fact it is the nervous system that is tense, this tenseness being communicated to skeletal muscles as low level electrical impulses resulting in muscle contraction.

Tension involves the quasi-persistent contraction of low threshold motor units. Within a given muscle, motor units are segregated into groups that possess differing sensitivities to nervous stimulation, smaller units being more sensitive than larger ones. Consequently, upon contraction of a given muscle, motor units are engaged and disengaged in a hierarchical order, smaller units "firing" at lower nervous potentials followed by larger units "firing" at higher nervous potentials. Upon relaxation, the reverse occurs. Therefore, smaller sensitive units are engaged first and remain engaged until the muscle is fully relaxed. It has been shown that smaller low threshold motor units may be activated by either physical activity or by psychological stress.[15] It may be assumed that, coincident with stress, the central nervous system generates the nerve potential necessary to initiate or sustain contraction – but how?

It is theorized that the tendency toward persistent tension, sustained contraction of small sensitive motor units, is a function of autonomic imbalance, specifically sympathetic emphasis. This emphasis serves to excite the entirety of the skeletal muscle system, either directly resulting in contraction of low threshold motor units or, at a minimum, biasing sensitive motor units such that they remain contracted when they would otherwise relax, the net effect being persistent muscle tension.

The effect of sympathetic action on low threshold motor units can be verified experientially. Hold out your dominant hand. Sense what it feels like. Now clench the other hand into a tight fist. Now sense your dominant hand. What do you feel? This exercise is employing previously discussed bridges, the hands being one of them. Try it a few more times if you wish, as it is good way to directly experience the action of the sympathetic nervous system that we have been discussing throughout this book. Now consciously relax for at least a minute.

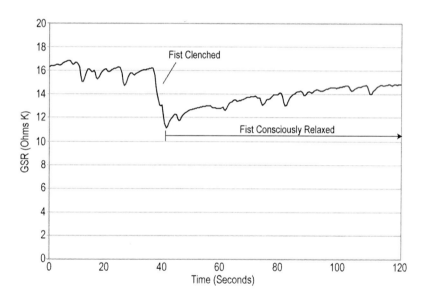

Figure 30: Galvanic skin response measured in left hand while clenching and then relaxing the right hand

Figure 30 presents a view of sympathetic nervous system bias as measured in the left hand using galvanic skin response while forming a fist with the right hand. Note that there is an immediate sympathetic nervous system response, body resistance values

The New Science of Breath

promptly dropping from 16000 ohms to approximately 11000 ohms upon clenching the fist.

We can feel this clearly in the hands because we have a highly developed sense of the hands. Now that you know what we are looking for, let's do another quick experiment. Sit up straight, relax, and try to develop a sense of your whole body. Now clench your fist again. Try to feel what is going on in your entire body. What do you feel? The fact is this. Clenching of the fist stimulates the entirety of the sympathetic nervous system, resulting in the contraction of low threshold muscle units throughout the entire body, in effect preparing you for fight or flight.

You may note from your own experience and from the above figure that while the sympathetic nervous system engages promptly upon stimulation, it takes much longer to disengage. As shown in Figure 30, while the sympathetic nervous system takes only a few seconds to engage, even after a minute resistance values have not returned to their initial levels. While each of the bridges exhibits slightly different behavior in this regard, this is in general the nature of the challenge when dealing with the sympathetic nervous system. It is fast to engage but slow and even stubborn to disengage. If you look back at Figure 4 depicting changes in sympathetic action in response to changing breathing frequency, you will note that the response is similar – the rate at which the sympathetic nervous system asserts emphasis is much faster than the rate of withdrawal. This tendency makes perfect sense in the light of the sympathetic nervous system's fight-flight mandate. During times of threat it is critical that the body react quickly and be prepared for sustained action. It is not critical that the body slow down rapidly when the threat no longer exists. *However, it is critical that in time it slows down completely.*

Gaining conscious control over sympathetic nervous system reaction and countering sympathetic bias via cultivation of parasympathetic influence are fundamental objectives of hatha yoga. Hatha yoga confronts this matter by requiring flexibility. As sympathetic bias is what prevents flexibility, the yogi must ultimately learn to overcome it. Because breathing is a fundamental determinant of sympathetic emphasis, it is something we do 24 hours a day, and because the breathing bridge is the most effective means of accessing the autonomic nervous system, optimal breathing is a requisite for successful hatha yoga. By employing the bridges as both thermometer and thermostat, the practitioner can both sense and govern autonomic nervous system behavior.

"Mindfulness" has long been accepted as a necessary condition for proper and effective yogic practice. As we now have an understanding of the relationship between somatic and autonomic nervous systems and the breathing bridge, this is understandable.

One cannot inhale deeply without somatic nervous system involvement. In other words, you cannot inhale deeply without some degree of intention. The diaphragm, the large sheath of muscle separating the thoracic and abdominal cavities, governs lung capacity. Like any other large muscle, while at rest or semi-activity the degree of contraction is governed by the degree of intent, i.e. little intent yields little contraction and large intent yields large contraction.

Therefore, without significant intent, breathing with depth is not possible. As intent is a function of mindfulness, mindfulness is required for depth. Recall from a prior discussion that depth relates strongly to frequency. Consequently, while not the primary value, a key reason why proper mindful breathing is "healthful" is because it functions to slow the breathing frequency and increase depth, once again balancing the autonomic nervous system and countering sympathetic dominance and its myriad effects, a condition that I theorize results largely from suboptimal breathing in the first place!

Taijiquan (tai chi) and qigong (chi gung), slow-motion practices of Taoist origin, are particularly interesting to consider in the light of autonomic balance. They facilitate mindful breathing by synchronizing the action of breathing with the actions being performed by the rest of the body. Based on the taiji paradigm, therapeutic exercises are designed very deliberately around the harmonious interplay of "yin" and "yang", the basic convention involving an ebb and flow to which both body movements and breathing are synchronized. Traditionally, these exercises are performed at a relatively slow pace, taijiquan or tai chi being the well-known example. It is well established that the practice of these arts enhances health, well-being, and longevity. While practices of Taoist origin contain many therapeutic qualities, slow conscious movement with synchronized breathing is a foundational aspect of them all.

Figure 31: The "Taiji Diagram" – representation of the taiji paradigm

The function of slow-motion movement is very interesting. Strenuous exercise strongly stimulates sympathetic function – merely thinking about strenuous exercise will cause skeletal muscles to tense slightly. It is known that even brisk passive movement of the arms and legs will stimulate breathing frequency due to excitement of proprioceptors in the limbs. Except for exactness of posture and motion, both tai chi and qi gong are ideally to be practiced in a state of utter and complete relaxation and "softness". Only then, as is said in the tai chi classics, will "the lighting of a fly set you in motion", this statement referring to the lightness and sensitivity that one is to cultivate via practice.

As a long-term student and practitioner of Taoist arts looking at it afresh, it seems quite clear that the system is ingeniously designed to exercise and strengthen the somatic nervous function while, as much as possible, avoiding sympathetic nervous system excitation. In the practice of tai chi, one aspires to make even the smallest movement of the body as deliberate as can be, and at the same time, with the exception of exacting posture and motion, the body is to remain as relaxed and still as possible. Another fascinating aspect of the practice is that motions that would typically require physical power to execute in a martial situation, for example "push", are executed coincident with exhalation (parasympathetic emphasis) and motions that typically would not require power, for example the withdrawal prior to "push", are executed coincident with inhalation (sympathetic emphasis). While exhaling coincident with strong exertion is very natural, it shows the degree to which the system of exercise is elaborately designed to maintain balance. Because one is to imagine that during powerful actions one is actually working against an external force, for example, pushing one's way through molding clay, the sympathetic nervous system does engage, but because one is simultaneously exhaling and relaxing deeply, autonomic balance is maintained. Over time, this practice yields profound physical power, speed, and sensitivity.

To summarize, exercises of Taoist origin are ingeniously designed to be practiced in the state of autonomic balance as well as to cultivate that state. Because the state of balance is the ground for meditation, this makes perfect sense.

Let us now turn to the consideration of mantra. As most practitioners of yoga or meditation are aware, the mantra is a word or phrase, typically of spiritual significance, that is recited over and over many times (the aspirant is encouraged to recite it at all possible times) for the purpose of evoking and maintaining a spiritual meditative state. Traditionally, a mantra is passed from guru to devotee. What mantra recitation does psycho-physiologically, and how it functions, has been an esoteric mystery for thousands of years. Having said as much, I am confident that a primary function of mantra recitation is the elicitation of autonomic nervous system balance and cardiopulmonary resonance.

There is a natural interval that governs audible exhalation and therefore speaking, chanting, and singing. We are all familiar with it because it influences the timing, the structure, and the content of what and how we communicate verbally as well as in written form. This interval literally defines our comfort zone relative to the length of spoken phrases and sentences, and is why we tend to structure speech as we do. Not surprisingly, this is also the natural interval of exhalation, this interval being 50 percent of the period of the intrinsic autonomic nervous system rhythm or ~6 seconds, the entire period being that of the *Fundamental Quiescent Rhythm*, i.e. ~12 seconds. Evidence of this can be seen throughout popular music, many songs having verses that are almost exactly 6 seconds in length. This interval is used throughout the music of the Beatles, a few examples being, I Wanna Hold Your Hand, She Loves You, and The Long and Winding Road, wherein either the harmony or the chorus employs the 6-second interval or a multiple thereof. A few other examples include Love Me Tender and I'm All Shook Up, by Elvis. If you either hum or sing along with the verses of these songs, guess what happens? You experience moments of cardiopulmonary synchrony and resonance. Extensive use of this interval could easily explain the overwhelming popularity of these two artists in particular – listening to and participating in their music makes you feel good and function better!

Mantra recitation performs the same function of eliciting autonomic nervous system balance and cardiopulmonary resonance. In fact, although not previously stated in these terms, it is proposed that this is its essential, original purpose.

The science of mantra has changed little in the last 3000-4000 years, its ancient beginnings dating back at least as early as the Vedas, the primary texts of Hinduism, circa 1500 B.C. Generally, traditional mantras have meaning and cadence, where cadence plays an essential role both in recitation and in developing the "psycho-physiological engram". It is proposed that a central function of mantra was, and still is, highly accurate and repeatable timing – in effect, a reference rhythm with which both the mind and the breathing cycle are synchronized. There are many mantras, some fitting this fundamental 6/12-second intrinsic autonomic nervous system interval and others not. For those that do, it is easy to understand their efficacy.

Chanting (audible mantra recitation) is typically a group activity. There is no question that something magical happens when a large number of like-minded people chant together. It is known that the electromagnetic field of the human organism is most powerful and pronounced during cardiopulmonary resonance. It is also known that the powerful resonant field of one person can affect the fields of others, leading them toward coherent synchrony. I have both experienced this and validated it biometrically. One explanation for the group phenomenon during chanting is that the electromagnetic fields of the many chanting individuals are pronounced and synchronized with each other, leading to large-scale group resonance and synchrony. Fascinating possibilities indeed!

Mantra	Inhalation (6-second interval)	Exhalation (6-second interval)
Om	Ooooooommmmmmmmmm	Ooooooommmmmmmmmm
Ham Sah	Hhhhhaaaaaaaammmmmm	Sssssaaaaaaaaaaahhhhhhh
Om Mani Padme Hum	Ommm Maanni Padme Huumm	Ommm Maanni Padme Huumm
Sat Nam	SaaaaTaaaaNaaaaMaaaa	SaaaaTaaaaNaaaaMaaaa

Figure 32: Examples of mantras and how they fit within the 6-second interval

Accuracy of chanting cadence can be reinforced via the use of a kriya, the best example possibly being "kirtan kriya" involving the sequential touching of thumb-forefinger, thumb-middle finger, thumb-ring finger, and thumb-little finger with each recitation. The mantra Sat Nam formally employs kirtan kriya, that is, they are practiced in unison. If each syllable and corresponding finger position is held for ~1.5 seconds, and if breathing occurs in synchrony, it elicits autonomic nervous system balance and cardiopulmonary resonance. If combined with relaxation and stillness, meditation naturally results.

"Mala japa", mantra recitation employing a "mala" (typically a string of 108 equally spaced beads) functions similarly, counting of the beads serving as a highly accurate and repeatable timing mechanism.

"Mudra" also plays a role in moderating autonomic nervous system balance. Again from *Light on Yoga*, "There is no asana

The New Science of Breath

like Siddha, no kumbhaka like Kevala, no mudra like Khechari, and no laya (absorption of the mind) like Nada."[16] Khechari mudra involves rolling the tongue backward and upward, stretching it upward so that the tip rests lightly against the roof of the nasal pharynx. In this position, the tongue functions as a pranic (energetic) bridge connecting psychic centers in the head with those in the throat and torso. During meditation, khechari mudra results in a substantially deeper meditative experience. It is hypothesized that it accomplishes this by eliciting parasympathetic emphasis and consequent autonomic balance.

While admittedly mysterious, a consequence of khechari mudra is that it naturally decreases breathing frequency and increases breathing depth. When the tongue is held in this position, it substantially reduces the capacity of the airway between the inner nasal orifice and the throat. It would seem that narrowing the nostrils would have the same effect, but it does not. If one leaves the tongue in place with the tip naturally touching the front teeth and pinches the nose slightly so as to reduce the volume of air passing through the nostrils, one immediately senses air deprivation. This sense of air deprivation does not occur during khechari mudra.

As with mantra, for khechari mudra to be powerful it must be cultivated via continued practice. The yoga aspirant is encouraged to hold khechari mudra, not only while meditating but throughout the day except when engaged in *necessary* speech. In this way, khechari mudra functions to slow the breathing frequency and increase breathing depth, leading toward autonomic balance and cardiopulmonary resonance. While mantras and mudras are in themselves a means of moderating autonomic nervous system balance, they are also tools for meditation. This is because meditation is a function of autonomic nervous system balance; this is to say that *meditation occurs in the state of balance.*

The Sanskrit term "nada" refers to "inner sounds" or sounds of the subtle body. "Absorption of the mind" refers to the yogi's concentration upon these inner sounds during times of contemplation and meditation. In the early stages of nada yoga, the yogi adopts "yoni mudra" (a.k.a. "sanmukhi mudra") wherein the thumbs are inserted into the ears and the fingers are used to symbolically seal the eyes, nose, and mouth, thereby isolating the yogi's senses from external worldly stimuli. Importantly, placing the thumbs in the ears and allowing the fingers to rest lightly upon the facial bones serves to both seal out external sounds and dramatically amplify internal sounds. In the initial stages of practice, there are two primary sounds to be heard, these being the sound of the breath and the sound of the heartbeat. Because the sound of the breath overpowers more subtle sounds, the aspirant is encouraged to breathe so as to silence the breath. When the breath is silenced, the sound of the heartbeat becomes prominent, and the more it is practiced, the more clearly it can be heard. *Yoni mudra is, in fact, an ancient form of heart rate variability biofeedback*, thus allowing the yogi to listen to his own heartbeat, thereby discerning the critical relationship between heart and breathing rhythms. Once this relationship is well understood and cultivated, there is no longer a need to listen as one is able to sense it clearly without listening.

Coherent Breathing results in autonomic nervous system balance. When combined with stillness and relaxation, autonomic nervous system balance results in meditation. The function of breathing frequency in Zen meditation is made clear

Figure 33: Yoni Mudra

The New Science of Breath

from this excerpt from *Zen and the Mind*, in which Tomio Hirai documents his ground-breaking investigation into the physiology of Zen meditation. "In meditation, as time passes and the breathing rate decelerates, abdominal breathing comes to predominate over thoracic breathing. Although investigations performed on twelve priests showed that not all of them reached a low level of four or five breaths per minute and that the more experienced the priest, the slower his breathing rate, all of the men tested breathed much slower during meditation than before it."[17] Describing the breathing pattern of a single Zen priest: "Before meditation, the rate is a normal seventeen or eighteen breaths a minute. At the beginning of meditation the rate decreases rapidly to remain at about four or five breaths per minute throughout meditation. At the end of meditation the rate increases to twenty or twenty-two breaths a minute but decreases to normal (seventeen or eighteen breaths per minute) before long."[18] Note that it is a premise of this *new science* that you do not just breathe slowly and deeply during meditation but at all times, circumstances permitting.

There exists a state between sleep and wakefulness. In the Tantras, this state is referred to as "pratyahara", the simplistic translation being "control over the senses". In Kashmir Shaivism, this state is referred to as "suksma-gati" meaning "subtle awareness". (There is some distinction between the Kashmiri and Tantric terms relative to specific subtle states. For now we will use the term pratyahara.) In both Tantric and Shaivist systems, this state is *the ground* for the practice of meditation and *the gate* to higher level yogic experiences.

In practice, pratyahara exists on the line between consciousness and unconsciousness. While in this state, without concentrative effort, one dips in and out of wakefulness. This is to say that, during the meditative experience, one is apt to have moments of awareness and loss-of-awareness. The objective is to remain awake and aware. For this reason, it is important to have an object of concentration, concentration providing an appropriate level of sympathetic stimulation. This object *must* include the breath, for keen concentration on the breath is the mechanism by which pratyahara is achieved, but may also include concentration on a psychic center, mantra, koan, nada, etc.

The New Science of Breath

Pratyahara has to be experienced to be apprehended. Entry is characterized by distinct energetic shifts, almost step-like in their sensation. There is a very clear sense of body homogeneity and with it a many-fold increase in the perception of the flow of internal energy. After some time, there is a distinct loss of body boundary and often a lightness as if the body is inflated with helium and lighter than air. At times the body may become immobile, with the inability to lift even a finger without strong effort. Upon cessation of the experience, one typically feels rejuvenated and free of mental and physical tensions. I will assert that this is in fact the case; that a function of this state, and of meditation in general, is the unraveling of mental and physical blocks and knots.

Coherent Breathing *results* in meditation, this being dependent on a few factors including concentration (on the breath), body position, relaxation, and stillness. Generally speaking, the more the body tends toward horizontal, the easier it is to access the state of pratyahara. On the other hand, the more horizontal the body is, the harder it is to remain awake and aware during the practice. This is also largely dictated by the autonomic nervous system, vertical body position resulting in sympathetic emphasis and horizontal posture resulting in parasympathetic emphasis. As autonomic nervous system balance is the objective during meditation, many may find that reclining at a 45-degree angle yields optimal results.

Returning for a moment to the subject of "kriya", in Sanskrit the term has a number of meanings including, "action", "labor", and "cure". Relative to yoga, "kriya" is commonly used to express two quite different but related meanings. The first and, as it is understood, the original use, refers to spontaneous physical movement resulting from the action of kundalini Shakti, the kriya being a physical manifestation of kundalini actively clearing and purifying the psychic channels (nadis). The second popular usage is in reference to physical motions performed within the context of kundalini yoga, a branch of yogic practice that has as its fundamental objective the awakening of kundalini Shakti. Kirtan kriya is an example of this type of kriya. The definition "cure" relates to the natural healing that often occurs as a consequence of the *spontaneous* kriya. Manifestation of the spontaneous kriya is generally considered to be an outcome of meditative practice, since meditation sets kundalini Shakti in motion. Likewise, the kriyas of kundalini yoga are held to be of a curative nature.

Just as Coherent Breathing results in meditation, it also results in the manifestation of spontaneous kriyas – as it makes sense that it would. Generally, this natural action is manifested as a consequence of exacting "resonant" posture, coherent (resonant) breathing, and mental focus on a psychic center or point.

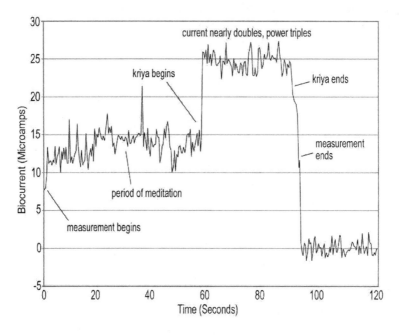

Figure 34: Bio-current of spontaneous "kriya" preceded by period of meditation employing Coherent Breathing

Figure 34 depicts a period of meditation employing Coherent Breathing, relaxation, and stillness, bio-current building over time consistent with Figure 28. In this case, when bio-current (measured to ground) approximates 15 microamperes, the kriya begins. It is sustained for approximately 40 seconds and is terminated intentionally for the purpose of presentation. It can be seen that during the period of the kriya, bio-current nearly doubles and power nearly triples.

The experience of kriya is a little difficult to describe and is manifested differently if you are seated rather than lying down, but it generally starts with a clear sense of throbbing or rhythmic undulation in the sushumna, or central channel (the chong meridian in the acupuncture system of China). When this rhythmic action is allowed to progress on its own, it will seemingly direct itself to an area of the body that is most in need of clearing or healing. Once one becomes very familiar with it, it can be consciously guided to any body part by bringing mental focus to that part, particularly a point of injury, pain, or discomfort, these points often relating to acupuncture points or "trigger points" in myofascial release parlance. When the action of the kriya is allowed to manifest itself fully, the entire body may become absorbed in a rhythmic twitching and jerking that can last for seconds, minutes, or even hours. It is held that "the action ends of its own accord when its work is complete".

Here again, the mechanism appears to be of an oscillatory nature. When both posture and breathing are resonant, and when energy is added to the system via mental focus, for example on a psychic center, energy builds until oscillation begins. While an exacting physiological explanation of what is going on here cannot be offered, it does seem apparent that it represents an innate capability of the organism to heal itself, as it does iron out knotted muscles, relieve pain, and promote circulation, nervous function, and comfort, even to areas of the body that have been compromised for many years – this, as well as enhancing vitality and overall psycho-physiological functioning. Having experienced both, it is proposed that "kriya" is in fact the same phenomenon that is known in myofascial release circles as "unwinding", a healing technique also known to "heal ancient wounds".[19]

The New Science of Breath

8

Coherent Breathing: The Practice

Recall, from the prior discussion, that Coherent Breathing can be facilitated in two basic ways. The first involves consciously synchronizing your breathing cycle with your heart rate variability cycle, and the second simply requires breathing at the rate of 5 cycles per minute, this frequency being essentially the same for all adults. When you breathe at this frequency, your heart rate variability cycle will synchronize with and "phase-lock" to your breathing cycle and they will remain synchronized for as long as they both remain regular and unperturbed.

Also recall that the sinewave represents ideal coherence, and for this reason it is a very useful model for Coherent Breathing – we want to breathe in a manner that approximates a sinewave. Here again, the pendulum is a useful analogy as it is a sinusoidal mechanical oscillator with which we are all familiar. If we plot 2 cycles of operation, we end up with 2 sinewaves as depicted by Figure 35.

Applying this analogy to the breathing cycle, exhalation begins at the uppermost peak of the wave and continues until the bottommost valley. Inhalation begins at the bottommost valley and continues until the uppermost peak. Just as the pendulum starts slowly, reaches maximal velocity at its midpoint and then naturally slows and changes direction, the breathing cycle also starts slowly, reaches maximal velocity at its midpoint and then naturally slows and changes direction.

The New Science of Breath

Given that our sinewave is 11.76 seconds in length, the period of the *Fundamental Quiescent Rhythm*, exhalation occurs over a period of 6 seconds (5.88 seconds to be precise) followed by inhalation for 6 seconds (5.88 seconds to be precise). It is our objective to breathe in a manner that emulates this action as closely as possible within reason. By this it is meant that the practice should be comfortable and natural and should *never* present significant strain.

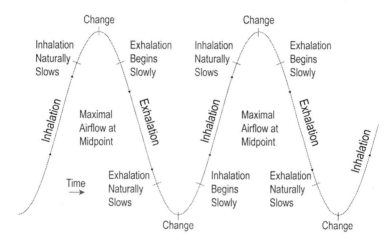

Figure 35: Sinewave model for Coherent Breathing

The RESPIRE 1 compact disc (visit www.coherence.com/ products) presents 3 different Breathing Pacemaker® audio recordings: "Vocal Instructive Sequence", "2 Bells", and "Clock & Bell". The Breathing Pacemaker is a simple audible or visual device that presents a reference rhythm for ideal breathing. You can think of it as a metronome for breathing. Although each pacemaker recording is somewhat unique, in some way each provides information pertaining to the beginning of the 6-second interval, the progression of the 6-second interval, and the end of the 6-second interval. In this way, the user is able to synchronize their breathing cycle with the recording with utmost accuracy and consistency. Some recordings distinguish between alternating 6-second intervals, i.e. one interval for exhalation and the next for inhalation, and others do not. This distinction is not particularly useful or important for a single user. Its purpose is to allow multiple users to synchronize their breathing cycles, thereby achieving group synchrony of autonomic nervous system rhythm.

Many people find it easiest to start with "Vocal Instructive Sequence" and then change once the understanding of what to do is clear. "Clock and Bell" is specifically designed to be used as an environmental background. It plays in my office at relatively low volume 24 hours a day. Because I also have a mechanical clock on the mantle, most people don't notice that it is not the clock that is ticking.

A common question regarding Coherent Breathing is: "How important is accuracy?" The answer to this question is rather lengthy and it has a lot to do with what you are trying to achieve.

Generally speaking, one should attempt to synchronize one's breathing cycle with that of the recording as closely as possible. Special attention should be given to synchronizing the moment of peak inhalation, the moment where inhalation ends and exhalation begins. This is because this is the moment in which the autonomic nervous system synchronizes to the breathing rhythm. If this moment occurs with regularity, this regularity will be translated into a regular "coherent" heart rate variability rhythm. Alternatively, if this moment occurs with irregularity, then the resulting HRV rhythm will be more irregular or "incoherent". As it is our objective that the HRV be as coherent as possible, it is important to pay close attention to this moment. Having said this, the practice should not require either mental or physical strain – it should be relatively easy and pleasant.

Accuracy, in general, translates into "coherence" – the higher the degree of accuracy (without strain), the higher the coherence of the intrinsic autonomic nervous system rhythm as reflected by the HRV cycle. Coherence is an extremely important factor in eliciting higher states of consciousness and capability as might be desired in order to accomplish any objective, for example healing. Remember that the rhythm we are speaking of is a primary electromagnetic rhythm of the human organism, permeating every cell and extending beyond the envelope of the physical body. Irregularity of this rhythm is likewise communicated to every cell.

The New Science of Breath

There are as many applications of Coherent Breathing as there are human endeavors. Having said this, no matter what your interest, it is important that you start at the beginning. So let's get started!

As we've discussed, the optimal breathing process approximates a sinewave as depicted in Figure 35. The process requires that we:

a) Breathe across the entire interval, inhaling for approximately 6 seconds and exhaling for approximately 6 seconds.

b) Change from inhalation to exhalation or vice versa at the specified moment,

c) Allow the pace of the air moving in and out of our lungs to naturally accelerate and decelerate, again very much like the action of the pendulum, and

d) Guide but do not force the process.

e) Eyes may be open or closed depending on your preference.

Visualize the sinewave as you breathe. Try to feel it throughout your body.

Inhalation requires effort just like flexing any other large muscle. Relax and let exhalation occur naturally of its own accord. It is important to remember that the breathing process is in itself an expression of yin and yang, non-action and action, decelerating and accelerating, relaxing and tensing – parasympathetic and sympathetic.

Exhalation elicits parasympathetic emphasis which begets relaxation; inhalation elicits sympathetic emphasis which begets tension. Recalling our discussion of "bridges", the sympathetic nervous function does not fully disengage without our conscious participation. Because exhalation is always preceded by inhalation, there is always residual sympathetic bias, this bias interfering with parasympathetic prominence. The result is that without conscious relaxation coincident with exhalation, it is often the case that the heartbeat rate does not decline to its natural limit. Of course, this results in a less than maximal HRV amplitude and persistent sympathetic bias. For this reason, conscious relaxation coincident with exhalation is an integral aspect of Coherent Breathing.

Stretching the breathing across the 6-second interval requires breathing to be slow and deep. For example, it is not correct to exhale for 3 seconds and hold for the remaining 3 seconds. Rather, the breath should be long and smooth, the air moving through your nostrils at a relaxed but continuous pace. During exhalation, as the lungs become empty and the diaphragm approaches full relaxation, the rate of exhalation will naturally slow. When I practice Coherent Breathing I cannot hear myself breathe.

Similarly, during inhalation, as the lungs become full and the diaphragm approaches full flexion, the rate of inhalation will naturally slow. Once the inhalation phase ends, simply allow exhalation to begin naturally and continue exhaling comfortably until the next cue.

Both inhalation and exhalation must be smooth and continuous. As breathing is governed by the diaphragm and intercostals, this requires that these muscles move smoothly, without irregularity. For most people, this is quite foreign. For this reason, Coherent Breathing requires toning of the diaphragm and intercostal muscle groups and related somatic nervous system control thereof. Therefore, it is best to set aside a short time each day for deliberate practice. With a little training, you will be able to control the motion of your diaphragm both fully and smoothly.

Adults tend to breathe at a rate between 10 and 20 breaths per minute with an average of 15, yielding 1 complete breathing cycle every 4 seconds. Unless you have been practicing a therapeutic breathing method, breathing at a rate 3 times less than this, i.e. 12 seconds per cycle or 5 breaths per minute, can prove challenging at first. Therefore, the initial challenge is learning to slow down the breathing frequency. "Deceleration" is a parasympathetic function. As such we begin the formal instructive method with our focus on exhalation.

While the following instructions are somewhat detailed, please rest assured that the method is quite simple. Having said this, for you to achieve maximal benefit it will be necessary for you to commit yourself to a fundamentally new approach to breathing.

Because Coherent Breathing is a process that brings about change in the human organism, both the practice and the experience change as it is practiced. Enjoy the journey!

Disclaimer:
Coherent Breathing is a physical breathing technique for the purpose of health enhancement. The activities, physical and otherwise, while not known to pose any danger, may, depending on individual health status, present a risk to some people. Those that choose to employ the method described herein should first consult a physician. As such, the author of this method and all accompanying documentation, together with his agents and his publisher, are not responsible in any manner whatsoever for any injury that may occur as a consequence of use or misuse of this product or its accompanying documentation.
THE METHOD AND ACCOMPANYING DOCUMENTATION IS FURNISHED "AS IS" AND WITHOUT WARRANTY AS TO MEDICAL ACCURACY OR EFFICACY. YOU ASSUME THE ENTIRE RISK AS TO THE RESULTS AND PERFORMANCE YOU MAY OBTAIN BY USE OR MISUSE OF THE METHOD AND ACCOMPANYING DOCUMENTATION. NO SPECIFIC CLAIMS ARE MADE REGARDING HEALTH BENEFITS.

Coherent Breathing – Formal Instructive Method:

This practice involves reorienting the autonomic nervous system. Approach it with patience. Practice only a few minutes per day at first. If at any time you feel pain or discomfort, promptly discontinue practice. The rhythm of 5 breathing cycles per minute applies to adults while in an erect body position and otherwise in a state of rest or semi-activity. It does not apply to aerobic exercise.

Step 1: Position yourself in a comfortable upright posture. Recline at no more than a 45-degree angle. Clothing should be comfortable and loose.

The New Science of Breath

Step 2: Start the RESPIRE 1 compact disc on your stereo or personal audio device and choose the recording of your liking. Some people find it easiest to begin with "Vocal Instructive Sequence".

Step 3: Consciously relax, attempting to let go of any tension you may be carrying.

Step 4: Listen closely to the pacemaker recording, noting the regularity of the cues to inhale and exhale.

Step 5: For the first few minutes of practice, every so often inhale deeply through your nose and then, timing your exhalation with the pacemaker, exhale for the duration of a single 6-second interval. Then resume your normal breathing rhythm. Don't attempt to do this continuously, just every so often. Do only what is comfortable for you.

Step 6: Once exhaling every so often for 6 seconds becomes familiar and comfortable, combine exhalation with deep relaxation. That is, just as in Step 5, every so often pick a 6-second audio interval and exhale for the duration while consciously relaxing and letting go of any stress or tension you may be holding. Employ the "bridges" as you do this. As you exhale, consciously focus your attention on relaxing the face, the perineum, the hands, and the feet. You may want to exhale through the mouth, relaxing your jaw and allowing it to drop of its own accord. You may also use a visual image to accentuate the relaxation process – the unwinding of a clock spring is a useful notion. Note any bodily sensations you may be experiencing.

In between exhalations, return to your normal breathing rhythm. With this as the focus of your practice session, continue for the time you have allotted. During moments throughout the day, exhale and relax. Use your Breathing Pacemaker recording to reinforce the timing.

Practice Steps 5 and 6 until you become very familiar and comfortable with the practice of exhalation. In time, you will note a distinct relaxation response that occurs coincident with each exhalation. (When exhalation occurs coincident with the parasympathetic phase of the heart rate variability cycle, a relaxation response naturally occurs. This sensation is indicative of effective parasympathetic action. A key reason that people

The New Science of Breath

become chronically tense is that this response is not elicited frequently. When you are breathing optimally, this response occurs with every exhalation!) When you have achieved this, proceed to Step 7 involving the incorporation of the 6-second inhalation interval.

(A note regarding breathing depth: Depth is an important factor in eliciting parasympathetic emphasis and consequent autonomic nervous system balance. Yet, unlike frequency, ideal depth is difficult to specify. Suffice it to say that there is a felt sense associated with ideal depth. While it should be completely comfortable, it does require slight effort. In time, you will feel a slight tingling as you approach full inhalation and a similar but different tingling when you approach full exhalation. This tingling signals that either inhalation or exhalation is nearing completion. For now, let's aim for the feeling of "comfortably deep" which may approximate to 70-80% of capacity.)

Step 7: Again, while listening to the Breathing Pacemaker recording of your choice, choose a 6-second interval and exhale for the duration. Listen for the next cue signaling the end of exhalation and the beginning of inhalation. (Each recording incorporates an indication that the cue is imminent, making transitions easy to predict.) When you hear the cue, begin inhaling gently through your nose. Continue inhaling smoothly and evenly until the next cue. You will notice the natural feeling that the lungs are filling and the breath is slowing down. You may also notice that inhaling for this duration requires conscious effort. Flexing the diaphragm to any significant extent requires conscious effort, just as flexing any other muscle of the body. This is a reason why optimal breathing must in fact be "mindful breathing". After each deliberate inhalation, relax and resume your regular breathing rhythm. Continue with this practice for the duration of your training period.

Again, during moments throughout the day, use your Breathing Pacemaker recording to reinforce this new inhalation rhythm. Inhale and exhale a few intervals at a time, then relax and resume your normal breathing rhythm. Note any body/mind changes that you may experience before and after a few cycles of breathing at the prescribed rhythm.

The New Science of Breath

Continue the practice outlined in Step 7 until you become comfortable and proficient.

Step 8: Now that you are proficient at both exhaling and inhaling for 6-second intervals, begin linking intervals together, first a single interval, then 2, then 3, etc. With this, your complete breathing cycle will be occurring on 12-second intervals yielding 5 complete breathing cycles in approximately 1 minute. Continue this practice of breathing at the 5 cycles per minute rate until you can do it comfortably for 1 minute, 2 minutes, 3 minutes, etc. Continue for as long as is comfortable without strain, then relax and return to your normal breathing rhythm.

Again, during moments throughout the day, use your Breathing Pacemaker recording to reinforce this new 5 cycles per minute rhythm.

Step 9: Continue with the practice of Step 8 until you are proficient in breathing at the 12-second interval for as often and for as long as you wish. During moments throughout the day, use your Breathing Pacemaker recording to reinforce this rhythm.

While at rest or semi-activity, make it your objective to breathe at this new rhythm all the time. There will be times that you will forget and revert to your regular unconscious breathing rhythm. In time, this will become noticeable and you will reorient your breathing in the moment to the new rhythm.

With practice, breathing at the target rhythm of 5 breathing cycles in ~1 minute will become commonplace for you. Having said this, modern life with its fast pace and many diversions tends to work against proper breathing. As previously mentioned, rapid shallow breathing is one of the first physical manifestations of stress. For this reason it is particularly important to note body/mind changes that occur when breathing at the target rhythm compared with those while breathing at the normal rhythm. These may include feelings of ease, relaxation, reduced anxiety, comfort, etc. In time, loss of these feelings will become reminders that you are not breathing at the optimal rhythm.

The New Science of Breath

9

Questions & Answers

1. What is Coherent Breathing?
Answer: Coherent Breathing involves breathing synchronously at the specific frequency of 1 cycle in approximately ~12 seconds with comfortable depth. Coherent Breathing results in autonomic nervous system balance, cardiopulmonary resonance, and coherence of the autonomic nervous system rhythm.

2. What is autonomic nervous system balance?
Answer: Autonomic nervous system balance is the state wherein sympathetic (activating) and parasympathetic (deactivating) functions of the autonomic nervous system are of equal emphasis.

3. Why should I care?
Answer: The emerging scientific and medical understanding is that many modern day maladies including anxiety, hypertension, attention deficiency, and chronic muscle tension are rooted in autonomic nervous system imbalance.

4. What is heart rate variability (HRV)?
Answer: HRV is the rate at which the heartbeat rate changes. It has the attributes of amplitude, frequency, average heartbeat rate, and coherence. Heart rate variability is reflective of autonomic nervous system status and correlates highly with health and well-being.

The New Science of Breath

5. What does the heart rate variability pattern mean?
Answer: Because HRV is a window on autonomic functioning, and because the autonomic nervous system underlies many aspects of the human organism, HRV is in many ways representative of the biostasis of the human organism.

6. What is cardiopulmonary resonance?
Answer: Cardiopulmonary resonance is the state wherein heart rate variability and breathing rhythms are synchronized and occurring at the frequency of resonance. Cardiopulmonary resonance is an outcome of autonomic nervous system balance.

7. What is coherence?
Answer: At the meta-level, coherence is representative of wholeness, communication and balance of the human organism. Relative to heart rate variability, coherence is a measure of the consistency of amplitude, phase, frequency, and function of the HRV rhythm.

8. What is sympathetic dominance?
Answer: Sympathetic dominance refers to a state of imbalance, specifically chronic over-emphasis of the sympathetic (activating) branch and under-emphasis of the parasympathetic (deactivating) branch of the autonomic nervous system.

9. How do you achieve autonomic nervous system balance?
Answer: Breathing *frequency* influences sympathetic emphasis and breathing *depth* influences parasympathetic emphasis. Consequently, there is a frequency and a depth where sympathetic and parasympathetic effects are equal. When one breathes at this frequency and depth, autonomic nervous system balance results.

10. What can I expect in the short term from the practice of Coherent Breathing?
Answer: Once you are employing Coherent Breathing during your day, each day you do, you can anticipate increased calm, comfort, and sense of well-being. *No specific claims are made regarding health benefits.*

11. What can I expect in the long term from the practice of Coherent Breathing?
Answer: You can expect the sense of increased calm and comfort to become lasting and continuous. You can also anticipate a shifting toward the psycho-physiological correlates of balance as described in Figure 21. *No specific claims are made regarding health benefits.*

12. What is the Breathing Pacemaker®?
Answer: The Breathing Pacemaker is a simple audio/visual device with which the user consciously synchronizes their breathing cycle for purposes of eliciting autonomic nervous system balance. It may be though of as a metronome for breathing.

13. Is it possible for a person to change how they breathe on an ongoing basis?
Answer: Yes, definitely. Modifying one's breathing is not unlike modifying one's posture. Once you know what to do and why you are doing it, it simply involves ongoing practice and commitment.

14. How long does it take to reprogram the breathing process?
Answer: In *PsychoCybernetics* (ISBN: 0-671-70075-8), Maxwell Maltz recommends practicing mindfully for 20 minutes per day for 21 days, after which the new engram will be established. Maltz also suggests that one reserve judgment regarding one's progress during this period.

15. Does the Breathing Pacemaker work on a subconscious or subliminal basis?
Answer: Yes, but only after the breathing has been thoroughly trained via conscious application and practice.

16. Is it necessary to pace both inhalation and exhalation?
Answer: Yes, inhalation and exhalation should be of equal length.

17. Is there a pause during the breathing rhythm, for example at the end of either inhalation or exhalation?
Answer: No. There is no pause in the heart rate variability rhythm and therefore no pause in the breathing cycle.

The New Science of Breath

18. Does the 12-second cycle apply to activities other than rest and semi-activity?
Answer: No. The 12-second cycle applies to times when the body is erect and otherwise at rest or semi-activity. As the body ramps up energy production coincident with physical activity, sympathetic nervous system action predominates and the frequency of the intrinsic autonomic nervous system rhythm increases dramatically.

19. Does the 12-second cycle apply when the body is resting horizontally?
Answer: While the 12-second rhythm may be used while horizontal, it is optimized for the vertical or erect body position. The autonomic nervous system understands and modifies sympathetic/parasympathetic emphasis with body inclination. As such, it increases sympathetic emphasis while vertical and parasympathetic emphasis while horizontal. For this reason, the intrinsic autonomic nervous system rhythm slows down while horizontal. Having said this, even while horizontal most adults breathe at a rhythm faster than 5 cycles per minute. For this reason, breathing at 5 cycles per minute while horizontal is beneficial. Additional Breathing Pacemaker products that are optimized for use while in a horizontal body position are available at: www.coherence.com.

20. Is the 12-second rhythm applicable to children?
Answer: While children may use the 12-second rhythm, it is not *optimized* for children. It is known that the frequency of resonance is higher in children than in adults, but how resonance changes with age has not been adequately characterized. Empirical evidence suggests that, by age 10, children demonstrate resonance in the .1 hertz range, approximately that of an adult. Use of the 12-second rhythm has proven effective when used by adolescents as part of an overall therapy program.

21. Can I use the Breathing Pacemaker during aerobic exercise?
Answer: No. See answer #18.

22. You recommend a breathing frequency of .085 Hz. or 5 breaths in approximately 1 minute. Is it not generally accepted that 0.1 Hz. is the frequency of resonance, .1 Hz. resulting in a breathing frequency of 6 breaths per minute?
Answer: My research indicates that *with toning*, the breathing frequency of .085 Hz. yields the highest HRV amplitude and is therefore the optimal frequency of resonance for adults in a state of rest or semi-activity and erect body position.

23. I am using my Breathing Pacemaker. How do I know if I am achieving autonomic balance and cardiopulmonary resonance?
Answer: A key sign of success is ease and comfort in breathing for minutes at a time at the approximate 12-second interval. If you are comfortable breathing at this frequency, a second key sign of success is this – if you sit quietly and breathe in the prescribed manner, within 2-3 minutes you will notice a distinct sense of internal quiet and comfort; you may start to feel a little sleepy. These feelings are more pronounced with eyes closed than with eyes open. What you are noticing is the new-found increase in parasympathetic (calm and relaxation) nervous system action. If and when you feel anxious throughout the day, can you employ this breathing method and reduce or eliminate these feelings? If so, then at a minimum you are being successful in consciously accessing reduced sympathetic and increased parasympathetic emphasis.

24. I have acquired a heart rate variability monitor and even though I am breathing at the rate of 5 cycles in 1 minute, I find that my HRV amplitude is not very high. What can I do to increase it?
Answer: While HRV amplitude is known to decline with age, *breathing frequency, depth,* and *relaxation* play a key part in moderating amplitude relative to its biological potential. In this situation, it is often the case that the HRV valley is higher than it might be if relaxing deeply upon exhalation. For this reason, emphasizing relaxation upon exhalation will often cause the HRV valley to drop lower, thereby increasing peak-to-peak HRV amplitude.

25. Does the use of caffeine effect autonomic nervous system balance?
Answer: Yes, caffeine stimulates sympathetic emphasis resulting in a relatively higher sympathetic bias. For this reason, autonomic nervous system balance may be difficult to elicit while under the influence of caffeine.

26. It is my understanding that autonomic nervous system balance is a goal of yoga and meditation. Is this true?
Answer: Yes. Successful yogic practice is predicated on autonomic nervous system balance. Therefore, yogic practices require the aspirant to understand and to overcome the root causes of imbalance.

27. Is the Breathing Pacemaker a tool used for practicing breathing or is it something you use all the time?
Answer: Both. The Breathing Pacemaker is to be used for deliberate breathing practice and as a background reference rhythm. You can think of the Breathing Pacemaker as a metronome for breathing.

28. Is cardiopulmonary resonance also possible when the body is active, for example during sport?
Answer: No. Technically, resonance exists at one frequency, the frequency of homeostasis, this depending on body position. Having said this, the heart rate variability rhythm will phase-lock with the breathing rhythm at any frequency if the breathing rhythm is synchronous.

29. Can I accomplish autonomic nervous system balance by breathing correctly for 20 minutes per day?
Answer: Practicing Coherent Breathing for 20 minutes per day will serve to shift your autonomic nervous system away from imbalance and toward balance, and for this reason it is a positive step toward improving one's well-being. However, it is far better to incorporate Coherent Breathing into your daily life such that imbalance does not occur in the first place.

30. *How long does autonomic nervous system balance last once achieved?*
Answer: For as long as your breathing remains optimally slow and deep.

31. *In practicing Coherent Breathing, do I need to be concerned about "abdominal" vs. "chest" breathing?*
Answer: No. If you breathe at the *Fundamental Quiescent Rhythm* with comfortable depth, the rest will take care of itself. This is to say, that a natural coordination between chest and abdomen will occur.

32. *Do I practice Coherent Breathing with my eyes open or closed?*
Answer: Both, depending on the circumstance and desired effect. Remember that the eyes are one of the mechanisms by which we consciously moderate autonomic function. Certainly it is easier to both concentrate and relax with eyes closed. At the same time, we want to learn to employ Coherent Breathing throughout the day with eyes open. Visual pacemaker products are available at www.coherence.com.

Appendix A

Brainwave Band	Frequency	Characteristics
High Beta	19-33 Hz.	hyper-alertness, anxiety, worry
Beta	15-18 Hz.	sustained attention-external orientation
SMR (low beta)	12-15 Hz.	mentally alert, physically relaxed
Alpha	8-12 Hz.	relaxed awareness; reverie
Theta	4-8 Hz.	meditation; half-waking, dreamlike imagery
Delta	0-4 Hz.	deep meditation; sleep

BCIA: Biofeedback Certification Institute of America (www.bcia.org)

Coherence: Relative to case observations, the HRV training program presents a "coherence figure" in %. This figure relates to the consistency of the heart rate variability cycle taking into account amplitude, frequency, and phase. For the purposes of the case discussions, this was generalized into 4 categories as defined below. These categories are referred to throughout the case observations where a specific percentage is not provided.

Very High Coherence ≥ 75%

50% ≥ High Coherence ≤ 75%

25% ≥ Moderate Coherence ≤ 50%

Low Coherence ≤ 25%

Electro-Cardiogram (EKG): An electrocardiogram is a graphic image produced by an electrocardiograph. An electrocardiograph records the electrical activity of the heart and is a major tool in cardiac electrophysiology.

Electro-Dermal Response (EDR): Electro-dermal response is a measure of current as a function of the impedance of the body to a weak electrical potential applied across two poles. It is measured in micromhos. EDR is the inverse of galvanic skin response (GSR). High EDR is indicative of sympathetic emphasis. Low EDR is indicative of sympathetic withdrawal.

Electro-Encephalograph (EEG): An instrument that detects, measures, and records electrical potentials on the scalp, these being representative of the electrical activity of the brain.

Galvanic Skin Response (GSR): Galvanic skin response is a measure of impedance of the body to a weak electrical potential applied across two poles. It is measured in ohms. GSR is the inverse of electro-dermal response. Low GSR is indicative of sympathetic emphasis. High GSR is indicative of sympathetic withdrawal.

Heart Rate Variability (HRV): Heart rate variability is the rate at which the heartbeat rate changes. HRV biofeedback involves monitoring and feeding back information pertaining to the heart rate variability cycle for the purpose of aiding a person in modifying his or her psycho-physiological state.

Neurotherapy: Neurotherapy, also "neurofeedback" and "EEG biofeedback", involves monitoring and feeding back information pertaining to "brainwaves" for the purpose of aiding a person in modifying his or her psycho-physiological state.

Neurophysiologic Assessment (NPA): Generally, neurophysiologic assessment involves the assessment of central nervous system state via electroencephalographic techniques.

Thermal Feedback: Thermal feedback simply involves interactive measurement of body temperature, usually of the extremities. Low temperature is indicative of sympathetic emphasis. High temperature is indicative of sympathetic withdrawal.

Test of Variables of Attention (TOVA): A continuous performance test assessing visual or auditory attention.

Appendix A

Figure 36: Pulse Amplitude and Blood Volume Measured At Cardiopulmonary Resonance

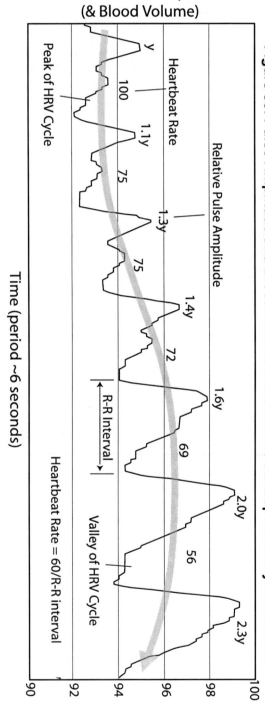

1. A plethysmograph measures changes in volume, in this case blood volume as measured in the finger.
2. Here we see the pulse wave superimposed on changing blood volume - *the respiratory arterial pressure wave.*
3. Pulse amplitude and blood volume are rising coincident with exhalation and falling heart rate.
4. Pulse amplitude and blood volume are 180 degrees out of phase with heart rate, i.e. maximal amplitude and volume occur at minimal heart rate and visa versa.

Appendix A

Figure 37: A Theory of Cardiopulmonary Operation at Resonance

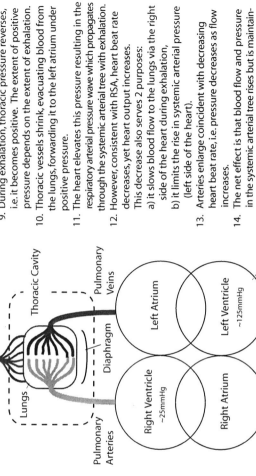

1. The pulmonary circulatory system holds 9% of body blood volume or about 450ml. However, it is capable of holding 2X its "normal" capacity.[20]
2. The pulmonary arterial tree has a very large compliance, equaling that of the entire systemic arterial tree.[21]
3. A function of the lungs and thoracic cavity is to serve as a reservoir supplying blood to the left atrium.
4. During inhalation, thoracic pressure becomes negative (a vacuum). Negative pressure increases with the extent of inhalation.[22]
5. Pulmonary blood vessels expand dramatically, storing blood, thereby reducing blood returning to the left atrium.[23] Systemic arterial pressure falls.
6. Consistent with RSA, heart beat rate increases, yet heart output decreases. This increase in heart beat rate serves 2 purposes:
   a) It ushers blood to the lungs via the right side of the heart during inhalation,
   b) It limits the fall in systemic arterial pressure (left side of the heart).
7. Arteries constrict coincident with increasing heart beat rate.
8. The net effect is that blood flow and pressure in the systemic arterial tree falls but is maintained within viable limits.
9. During exhalation, thoracic pressure reverses, i.e. it becomes positive. The extent of positive pressure depends on the extent of exhalation.
10. Thoracic vessels shrink, evacuating blood from the lungs, forwarding it to the left atrium under positive pressure.
11. The heart elevates this pressure resulting in the respiratory arterial pressure wave which propagates through the systemic arterial tree with exhalation.
12. However, consistent with RSA, heart beat rate decreases, yet heart output increases. This decrease also serves 2 purposes:
    a) it slows blood flow to the lungs via the right side of the heart during exhalation,
    b) it limits the rise in systemic arterial pressure (left side of the heart).
13. Arteries enlarge coincident with decreasing heart beat rate, i.e. pressure decreases as flow increases.
14. The net effect is that blood flow and pressure in the systemic arterial tree rises but is maintained within viable limits.
15. In this way, the heart, lungs, and cardiovascular system work in unison to facilitate the arterial pressure wave which rises with exhalation and falls with inhalation.
16. This process is powered by the action of the diaphragm and intercostals.

Appendix A

Figure 38: Thoracic Cavity - Source & Sink

Bronchial arteries from systemic circulation (1-2% of total cardiac output)

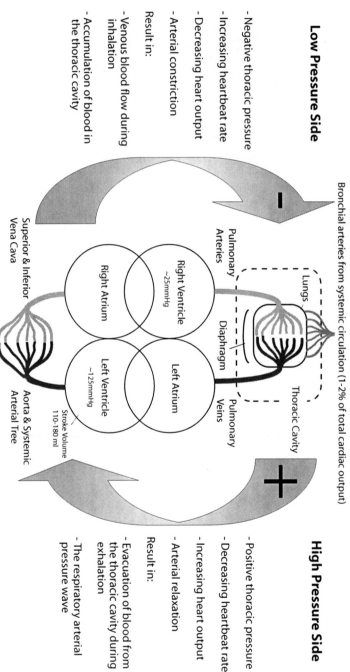

Low Pressure Side

- Negative thoracic pressure
- Increasing heartbeat rate
- Decreasing heart output
- Arterial constriction

Result in:
- Venous blood flow during inhalation
- Accumulation of blood in the thoracic cavity

High Pressure Side

- Positive thoracic pressure
- Decreasing heartbeat rate
- Increasing heart output
- Arterial relaxation

Result in:
- Evacuation of blood from the thoracic cavity during exhalation
- The respiratory arterial pressure wave

Appendix A

Figure 39: Mechanics of Cardiopulmonary Resonance - Schematic View

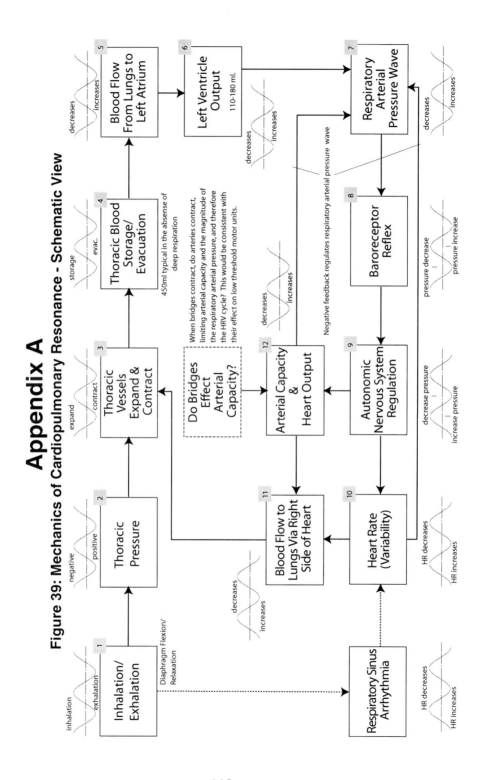

Notes

1. Reprinted from *The Primordial Breath*, Huang and Wurmbrand, p. 12, © 1987, with permission of Original Books Inc. (xxi)

2. Reprinted from UCSF, John and Catherine MacArthur Research Network, *Parasympathetic Function*, McEwen, Bulloch, and Stewart, © 1999, with permission of the University of California – San Francisco. (11)

3. Reprinted from *Medical Physiology*, Guyton & Hall, p. 193, © 2000, with permission from Elsevier. (11)

4. Reprinted from *Medical Physiology*, Guyton & Hall, p. 193, © 2000, with permission from Elsevier. (17)

5. Vaschillo, et al. (2002), p. 21. (17)

6. Guyton & Hall, (2002), p. 193. (17)

7. *Cardiorespiratory and Cardiosomatic Psychophysiology*, Grossman, Janssen, & Vaitl, Editors. NATO ASI Series (1983) (20)

8. Kitney (1986), p. 83. (20)

9. Reprinted from *The Rainbow and the Worm*, Ho, page 229, © 2003, with permission of Mae-Wan Ho (www.i-sis.org.uk) and World Scientific Publishing. (27)

10. Childre, Martin (1999) p. 40. (30)

11. Norretranders (1991), p. 139. (40)

12. Reprinted from *American Heart Journal*, De Meersman, Volume 125: 726-731, with permission from Elsevier. (41)

13. Reprinted from *Light on Yoga*, Iyengar, p. 45, by permission of HarperCollins Publishers Ltd., © B.K.S. Iyengar, 1966. (77)

14. Reprinted from *Light on Yoga*, Iyengar, p. 46, by permission of HarperCollins Publishers Ltd., © B.K.S. Iyengar, 1966. (78)

15. UCSF, John and Catherine MacArthur Research Network, *Allopathic Load Working Group*, McEwen, Bulloch, and Stewart,

© 1999, with permission of the University of California – San Francisco. (78)

16. Reprinted from *Light on Yoga*, Iyengar, p. 91, by permission of HarperCollins Publishers Ltd., © B.K.S. Iyengar, 1966. (86)

17. Reprinted from *Zen and the Mind*, Hirai, p. 113, © 1978, by permission of Japan Publications Inc. (88)

18. Reprinted from *Zen and the Mind*, Hirai, p. 113, © 1978, by permission of Japan Publications Inc. (88)

19. Barnes (2000). (91)

20. Guyton, A., Hall, J., *Medical Physiology*, W. B. Saunders, Philadelphia, p. 445. 2000 (109)

21. Guyton, A., Hall, J., *Medical Physiology*, W. B. Saunders, Philadelphia, p. 444. 2000 (109)

22. Guyton, A., Hall, J., *Medical Physiology*, W. B. Saunders, Philadelphia, p. 193. 2000 (109)

23. Guyton, A., Hall, J., *Medical Physiology*, W. B. Saunders, Philadelphia, p. 193. 2000 (109)

References

Applied Psychophysiology and Biofeedback (AAPB), www.aapb.org.

Apter, M.J., *Reversal Theory*, Routledge, London, 1989.

Antelmi, I., Silva De Paula, R.S., Shinzato, A.R., Peres, C.A., Mansur, A.J., Grupi, C.J., *Influence of Age, Gender, Body Mass Index, and Functional Capacity on Heart Rate Variability in a Cohort of Subjects Without Heart Disease*, American Journal of Cardiology, February 2004, pp. 381-385.

Barnes, J.F., *Healing Ancient Wounds*, 1st ed., MFR Treatment Centers & Seminars, Lancaster, 2000.

Bass, G., *Nonlinear Man – Chaos, Fractal & Homeostatic Interplay in Human Physiology*, 1997, www.dchaos.com

Becker, R.O., Selden, G., *The Body Electric*, William Morrow & Company, New York, 1985.

Bernardi, L., Wdowczyk-Szule, J., Valenti, C., Castoldi, S., Passino, C., Spadacini, G., Sleight, P., *Effects of Controlled Breathing, Mental Activity and Mental Stress With or Without Verbalization on Heart Rate Variability*, Journal of American College of Cardiology, Vol. 35, No. 6, 2000.

Berntson, G., Bigger, T., Eckberg, D., Grossman, P., Kaufman, P., Malik, M., Nagaraja, H., Porges, S., Saul, J., Stone, P., Van Der Molen, M., *Heart Rate Variability: Origins, Methods, and Interpretive Caveats*, Psychophysiology, Cambridge University Press, No. 34, 1997.

Biagini, M., Cammarota, C., Prisco, M., Di Liberato, F., Fiori, V., Greziosi, P., Perelli, P., Romano, R., Lanza, M., *Autonomic Nervous System Function Assessed By Analysis of Heart Rate Variability At Rest and During Exercise In Hypertensive and Normotensive Subjects*, American Journal of Hypertension, Volume 17, Issue 5, Supplement 1, May 2004.

Cade, C.M., Coxhead, N., *The Awakened Mind*, Delacorte Press, New York, 1979.

Cammann, H., Michel, J., *How to Avoid Misinterpretation of Heart Rate Variability Power Spectra*, Computer Methods and Programs in Biomedicine, 68, 2002.

Cheng, Man-Ching, *Master Cheng's Thirteen Chapters on Tai Chi Chuan*, Sweet Chi Press, New York, 1982.

Cherniske, S., *The DHEA Breakthrough*, Ballantine, New York, 1996.

Childre, D., Martin, H., *The Heartmath Solution*, Harper-Collins, 1999.

Childre, D., McCraty, R., *Psychophysiological Correlates of Spiritual Experience*, Biofeedback (Winter 2001).

Cleary, T., *The Secret of the Golden Flower*, Shambala, Boston, 1991.

Cleary, T., *Vitality Energy Spirit*, Shambala, Boston, 1991.

Csikszentmihalyi, M., *Flow – The Psychology of Optimal Experience*, HarperCollins, New York, 1990.

De Meersman, R., *Heart Rate Variability and Aerobic Fitness*, American Heart Journal, Volume 125, Number 3, March 1993.

Douglass, W.C., *Stop Aging or Slow the Process*, Rhino Publishing, Panama, 2003.

Feuerstein, G., *The Shambala Guide to Yoga*, 1st ed., Shambala, Boston, 1996.

Fox, E.L., Bowers, R.W., Foss, M.L., *The Physiological Basis for Exercise and Sport*, 5th ed., Brown & Benchmark, Madison, 1993.

Kitney, R.I., *Heart Rate Variability in Normal Adults*, Grossman, P., Janssen, K.H.L., Vaitl, D. editors. *Cardiorespiratory and Cardiosomatic Psychophysiology*; Plenum Press, New York, 1986, in cooperation with NATO Scientific Affairs, pp. 83-100.

Grant, G., Parr, T., *The Decline of Life's Energy Theory of Aging*, ati.uta.edu/DOLE2.pdf

Guyton, A., Hall, J., *Medical Physiology*, W.B. Saunders, Philadelphia.

Hirai, T., *Zen and the Mind*, Japan Publications Inc., Tokyo, 1978.

Ho, Mae-Wan, *The Rainbow and The Worm – The Physics of Organisms,* 2nd ed., World Scientific Publishing Company Pte. Ltd., Singapore, 2003.

Huang, J., Wurmbrand, M., *The Primordial Breath* – Volume I, Original Books, Inc., Torracne, CA. 1987.

Hughes, J., *Self Realization in Kashmir Shaivism*, State University of New York Press, 1995.

International Society for Neuronal Regulation, www.isnr.org

Iyengar, B.K.S., *Light On Yoga*, Schocken Books, New York, 1976.

Jensen, R.J., Schultz, G.W., Bangerter, B.L., *Applied Kinesiology and Biomechanics*, 3rd ed., McGraw Hill, 1993.

Kohn, L., *Taoist Meditation and Longevity Techniques*, Center for Chinese Studies, University of Michigan, 1989.

Lakshmanjoo, Swami, Hughes, J. editor, *Kashmir Shaivism – The Secret Supreme*, Kashmir Shaivism Fellowship, 2000.

Levine, P.A., *Waking the Tiger – Healing Trauma*, North Atlantic Books, Berkeley, 1997.

Lewis, D., *The Tao of Natural Breathing*, Mountain Wind Publishing, San Francisco, 1997.

Liao, W., *Tai Chi Classics*, Shambala Publications, Boston, 1977.

MacArthur, J., MacArthur, C., *Research Network on Socioeconomic Status and Health*, Parasympathetic Function, (www.macses.ucsf.edu/research/allostatic/notebook/parasym.html), Summary prepared by McEwen, B., Bulloch, K., and Stewart, J., July 1999.

MacArthur, J., MacArthur, C., *Research Network on Socioeconomic Status and Health*, Allostatic Load Working Group, (www.macses.ucsf.edu/research/allostatic/notebook/heart.rate.html), Summary prepared by McEwen, B., Bulloch, K., and Stewart, J., July 1999.

Macnaughton, I., *Body, Breath, and Consciousness*, North Atlantic Books, 2004.

Malik, M., Chairman, *Heart Rate Variability*, Writing Committee of the Task Force, Task Force of the European Society of Cardiology and the North American Society of Pacing and Electrophysiology, European Heart Journal, No. 17, 1996.

Maltz, M., *Psychocybernetics*, Pocket Books, New York, 1960.

Mandelbrot, B. B., *The Fractal Geometry of Nature*, W.H. Freeman and Company, New York, 1977.

Muktananda, Swami, *Play of Consciousness*, 3rd ed., SYDA Foundation, South Fallsburg, 2000.

Norretranders, T., *The User Illusion*, Penguin Putnam, Inc., New York, 1991.

Oschman, J., *Energy Medicine in Therapeutics and Human Performance*, Elsevier Limited, 2003.

Painter, J., *Flying Dragon Qi Gong*, IAM, 2005.

Pikkujamsa, Sirkku, *Heart Rate Variability and Baroreflex Sensitivity in Subjects Without Heart Disease*, Department of Internal Medicine, Oulu, 1999.

Pumpria, J., Howorka, K., Groves, D., Chester, M., Nolan, J., *Functional Assessment of Heart Rate Variability: Physiological Basis and Practical Applications*, International Journal of Cardiology 84 (2002).

Prabhakara, S., Krishnan, S., Banerjee, S., *Analysis of Human Heart Conditions Using Non-Linear Dynamics*, National Conference on Non-linear Systems and Dynamics, Indian Institute of Technology, Kharagpur, 2003.

Ramacharaka, Yogi, *Science of Breath*, Yogi Publication Society, Chicago, 1905.

Ramacharaka, Yogi, *Hatha Yoga*, Yogi Publication Society, Chicago, 1930.

Saraswati, Swami Satyananda, *Meditations from the Tantras*, 2nd ed., Bihar School of Yoga, Bihar, 1983.

Schafer, C., Rosenblum, M., Abel, H., Kurths, J., *Synchronization In The Human Cardiopulmonary System*, Physical Review E, Vol. 60, Number 1, The American Physical Society, 1999.

SenSharma, D.B., *The Philosophy of Sadhana*, State University of New York Press, 1990.

Shapiro, B.A., Kacmarek, R.M., Cane, R.D., Peruzzi, W.T., Hauptman, D., *Clinical Application of Respiratory Care*, 4th ed., Mosby Year Book, St. Louis, 1991.

Sime, W. E., *Psychophysiology of Stress and Relaxation*, Department of Health and Human Performance, University of Nebraska-Lincoln,
www.unl.edu/stress/management/psychophys.htm

Thompson, M., Thompson, L., *The Neurofeedback Book*, The Association of Applied Psychophysiology and Biofeedback, 2003.

Todd, M.E., *The Thinking Body*, Paul B. Hoeber, Inc., 1937.

Tripathi, K., *Respiration and Heart Rate Variability: A Review With Special Reference To Its Application In Aerospace Medicine*, Industry Journal of Aerospace Medicine, 48(1): 64-75, 2004.

Yasuma, F., Hayano, J., *Respiratory Sinus Arrhythmia: Why Does the Heartbeat Synchronize with Respiratory Rhythm? – Opinions, Hypotheses*, Chest –The Cardiopulmonary and Critical Care Journal, Feb. 2004.

Vaschillo, E., Vaschillo, B., Lehrer, P., *Heartbeat Synchronizes With Respiratory Rhythm Only Under Specific Circumstances*, Chest – The Cardiopulmonary and Critical Care Journal, 126: 1385-1387, 2004.

Vaschillo, E., Lehrer, P., Rishe, N., Konstantinov, M., Heart Rate Variability Biofeedback As A Method For Assessing Baroreflex Function: A Preliminary Study of Resonance In The Cardiovascular System, *Applied Psychophysiology and Biofeedback*, Vol. 27, No. 1, 1-27 (2002).

Vaschillo, E., Vaschillo, B., Lehrer, P., Characteristics of Resonance In Heart Rate Variability Stimulated by Biofeedback, *Applied Psychophysiology and Biofeedback*, Vol. 31, No. 2, 129-142 (2006).

Wise, A., *Awakening the Mind*, Tarcher/Putnam, New York, 2002.

Index

9 Dragon Baguazhang.. 27

abdominal breathing... 88
accuracy..85, 93, 94, 97
ADD... 44, 53, 67, 68
addiction....................44, 49, 50, 59, 60, 61, 62, 74, 76, 119
ADHD... 44, 53
adult breathing... 4, 11
aging... 7, 41, 113
allostatic load.. 11, 114
alpha................................... 45, 52, 59, 60, 62, 71, 72, 74, 75, 108
alpha/theta.. 52, 60, 62, 71, 75
alpha/theta training... 52, 60, 71
amplitude.. 9, 10, 11, 12, 13, 14, 15, 17, 21, 24, 29, 30, 31, 34 35, 36, 37, 38, 41, 42, 44, 45, 47, 48, 49, 51, 52, 54, 55, 56, 57 58, 59, 60, 62, 63, 64, 66, 68, 69, 71, 72, 73, 74, 95, 101, 102 105, 106, 108
anal sphincter..3
anger... 50, 51, 52, 74
anxiety........... 6, 7, 32, 45, 46, 47, 49, 50, 51, 54, 61, 62, 63, 65 67, 68, 69, 70, 71, 74, 100, 101, 108, 119
arterial pressure...17, 18, 24, 25, 26, 36, 38
arterial pressure wave................................ 18, 25, 26, 36, 37, 38
attention deficiency.. 7, 101
autonomic nervous system... 1, 2, 3, 5, 6, 7, 8, 9, 10, 17, 18, 19 21, 22, 24, 25, 27, 28, 29, 30, 31, 34, 36, 37, 39, 74, 76, 77, 78 81, 83, 84, 85, 86, 87, 89, 94, 97, 99, 101, 102, 103, 104, 106 107, 112
autonomic nervous system balance... 16, 22, 28, 37, 39, 74, 76 77, 78 83, 84, 85, 86, 87, 89, 99, 101, 101, 102, 103, 106, 107
autonomic nervous system governance.........................5
autonomic subsystem... 1

baroreceptor...36
beta........... 45, 51, 52, 53, 56, 57, 58, 59, 60, 62, 63, 64, 67, 68 70, 71, 75, 108

bioenergy.. 38, 39, 41, 43, 90
biological age.. 41, 42
biometrics.. 9, 44, 45, 46, 70
blood..18, 24, 25, 26, 36, 38
blood flow...17, 25, 36
blood oxygen...40
blood pressure...6, 7, 9, 18, 26, 49, 50
blood stagnation...26, 38
blood sugar..59, 61
blood vessels...17, 24, 25, 36
body inclination.. 35, 104
breath.....1, 3, 5, 42, 46, 47, 75, 76, 78, 87, 88, 89, 96, 99, 110 114, 115, 120
breathing.... 1, 2, 3, 4, 5, 6, 8, 11, 12, 13, 14, 15, 16, 17, 18, 21 22, 23, 24, 26, 30, 31, 32, 33, 34, 35, 36, 38, 41, 42, 43, 44, 45 47, 48, 50, 51, 52, 53, 54, 55, 57, 58, 59, 62, 63, 64, 66, 70, 71 72, 73, 74, 75, 76, 77, 78, 80, 81, 82, 84, 85, 87, 88, 90, 92, 93 94, 95, 96, 97, 98, 99, 100, 101, 102, 103, 104, 105, 106, 107 114, 119
breathing bridge...3, 4, 19, 31, 81
breathing depth..4, 19, 35, 36, 41, 86, 99, 102
breathing frequency.............................. 17, 19, 24, 31, 35, 36, 42
Breathing Pacemaker.... 44, 93, 98, 99, 100, 103, 104, 105, 106
breathing rate.. 88
bridge... 3, 86
bronchi... 24

caffeine... 49, 106
carbon dioxide... 1, 12, 24, 25, 37, 38
cardiac.. 10, 25, 65, 108
cardiac sudden death... 11
cardiopulmonary resonance.........20, 26, 30, 32, 33, 36, 38, 77 83, 84, 85, 86, 101, 102, 105, 106
cardiovascular constriction... 18
ceiling... 35, 41
central nervous system............................. 1, 17, 20, 42, 78, 109
chanting... 83, 84, 85
chaos... 30, 112

circulation..38
coherence........... 9, 10, 11, 12, 13, 14, 27, 29, 30, 38, 41, 42, 44
45, 47, 48, 51, 52, 53, 54, 55, 56, 57, 58, 59, 60, 61, 62, 63, 64
65 66, 68, 69, 71, 72, 73, 74, 92, 94, 101, 102, 104, 107, 108
Coherent Breathing... 31, 33, 39, 43, 44, 45, 46, 47, 48, 49, 50
51, 52, 53, 54, 55, 56, 57, 58, 59, 60, 61, 62, 63, 64, 65, 66, 67
68, 69, 70, 71, 72, 73, 74, 75, 76, 77, 87, 89, 90, 91, 92, 93, 94
95, 96, 97, 101, 102, 103, 106, 107
communication.. 1, 27, 29, 30, 102
concentration.. 59, 68, 78, 86, 88, 89
conscious............... 2, 3, 7, 18, 19, 29, 32, 40, 54, 63, 75, 78, 80
82, 95, 96, 99, 103
conscious participation.. 7, 29, 95
consciousness.. 7, 40, 41, 88, 94, 115
coronary heart disease.. 11
cranial respiratory impulse.. 21

deep respiration.. 11, 18, 25
delta........... 35, 51, 52, 53, 56, 57, 58, 59, 63, 64, 67, 68, 75, 108
delta/theta................ 51, 52, 53, 56, 57, 58, 59, 63, 64, 67, 68, 75
depression...... 44, 46, 61, 62, 67, 68, 69, 70, 71, 72, 73, 74, 119
diaphragm.. 3, 4, 81, 96, 99
Dr. Elsa Baehr.. 45
Dr. Roger Riss.. 46
Dr. Ronald DeMeersman...40, 41
dual control.. 3, 4, 5, 6

EDR........................... 46, 47, 48, 49, 64, 70, 71, 75, 109
EEG........... 45, 46, 47, 48, 49, 51, 52, 53, 56, 57, 58, 59, 60, 62
63, 64, 67, 68, 69, 70, 71, 72, 73, 74, 75, 109
EKG.. 65, 108
electro-cardiogram.. 108
electro-dermal response........................... 45, 46, 63, 64, 109
electroencephalograph.. 30
emotions.. 32, 56, 74
entropy.. 27, 30
excretory..3

exhalation............ 11, 18, 24, 31, 33, 34, 35, 38, 42, 50, 83, 84
92, 93, 94, 95, 96, 98, 99, 103, 105
eyes 3, 5, 42, 48, 49, 52, 54, 58, 59, 60, 70, 71, 73, 87, 95
105, 107

feet... 3, 5, 42, 68, 98
fight or flight.. 8, 80
floor.. 5, 35, 41, 42
fast Fourier transformation.. 10
frequency.... 4, 5, 6, 8, 9, 10, 11, 14, 17, 19, 20, 21, 22, 23, 24
29, 30, 31, 33, 34, 35, 36, 37, 42, 44, 45, 46, 60, 65, 67, 68, 75
80, 81, 82, 86, 87, 92, 96, 99, 101, 102, 104, 105, 106, 108
Fundamental Quiescent Rhythm........21, 31, 32, 34, 41, 84, 93, 107

galvanic skin response.. 5, 30, 79, 109
gas exchange.. 1, 34
GSR... 4, 109

hands................................... 3, 5, 6, 30, 42, 68, 79, 80, 98
healing...................... 43, 89, 90, 91, 94, 112, 114, 119
heart output.. 18, 36
heart rate variability... 8, 9, 10, 11, 15, 16, 17, 18, 21, 22, 24, 25
27, 28, 29, 30, 31, 33, 34, 36, 37, 38, 41, 42, 44, 45, 46, 47, 49, 51
53, 54, 65, 69, 71, 74, 87, 92, 94, 98, 101, 102, 104, 105, 106, 108
109, 112, 113, 115, 116
heartbeat rate.................. 8, 9, 10, 11, 12, 13, 14, 15, 17, 18, 24 31
33, 34, 35, 36, 68, 95, 101, 109
Helmuth Frank.. 40
high blood pressure...7
Ho, Mae-Wan...27
homeostasis.. 6, 30, 31, 34, 106
HRV...... 8, 9, 10, 11, 12, 13, 14, 15, 16, 21, 22, 23, 24, 25, 30, 33
34, 35, 36, 37, 38, 40, 41, 42, 44, 45, 46, 47, 48, 49, 51, 52, 53, 54
55, 56, 57, 58, 59, 60, 62, 63, 64, 65, 66, 67, 68, 69, 70, 71, 72, 73
74, 94, 95, 101, 102, 105, 106, 108, 109
HRV(av).. 10
human organism................... 1, 11, 27, 29, 43, 84, 94, 97, 102
human physiological capacity.. 40
hypertension.. 18, 44, 49, 74, 101, 112

imbalance.................. 7, 8, 41, 77, 78, 101, 102, 106, 107
incoherence.. 27, 30
inhalation......... 11, 18, 24, 31, 33, 34, 35, 36, 38, 50, 57, 83, 92
93, 94, 95, 96, 99, 103
instantaneous arterial pressure.. 17, 18
internal energy...89
intrinsic autonomic nervous system rhythm..... 21, 30, 84, 94, 104

jaw... 3, 5, 42, 98
Kashmir Shaivism.. 88, 114
khechari mudra... 86
kidneys... 7
kirtan kriya... 85, 89
kriya.. 39, 85, 89, 90, 91
kundalini.. 89, 90
kundalini yoga.. 89, 90

Lamaze... 44
lungs.............................. 10, 24, 35, 36, 37, 38, 95, 96, 99

mala japa...85
mantra.. 83, 84, 85, 86, 88
martial art... 42
martial arts... 119
meditation...77, 83, 85, 86, 87, 88, 89, 90, 91, 106, 108, 114, 119
mind... 15, 19, 27, 28, 29, 32, 33, 38, 40, 43, 49, 50, 55, 61, 67, 70
76, 78, 84, 86, 88, 99, 100, 111, 113, 114, 116, 119
mindfulness... 19, 81
mortality.. 11, 42
mudra.. 85, 86
muscle groups.. 3, 4, 30, 96
muscle tension... 6, 7, 8, 32, 79, 101
myofascial release... 90, 91

nada.. 86, 88
Neuro-Physiologic Assessment.. 46, 109
neurotherapy...... 45, 47, 53, 56, 60, 69, 71, 72, 74, 76, 109, 119
NPA.. 46, 47, 51, 53, 59, 61, 67, 70, 109

obsessive compulsive disorder (OCD) 44
oscillator... 20, 21, 24, 29, 30, 92
oxygen... 1, 12, 24, 25, 38, 40

pain................................. 7, 49, 50, 74, 76, 90, 91, 97
panic attack.. 65, 66, 74
parasympathetic... 2, 4, 5, 7, 9, 10, 14, 15, 16, 17, 18, 22, 30, 31 32, 34, 36, 37, 78, 80, 83, 86, 89, 95, 96, 98, 99, 101, 102, 104 105, 110, 114
parasympathetic emphasis........... 4, 5, 14, 15, 16, 17, 30, 31, 34 36, 83, 86, 89, 95, 99, 102, 104, 105
peak................ 10, 15, 16, 22, 26, 35, 36, 37, 40, 41, 92, 94, 106
pendulum.. 24, 25, 92, 95
performance anxiety..................................... 44, 47
perineum... 5, 42, 98
pH.. 25
phase................ 20, 24, 25, 29, 30, 31, 92, 96, 98, 102, 106, 108
phase-lock... 20, 24, 92, 106
phase shift... 17
plesthymograph.. 18
pranayama.. 33, 77, 78
pratyahara.. 78, 88, 89
psycho-physiological correlates..................... 28, 103

Q.. 21
qi.. 26, 43, 82, 115
qigong... 81
quantum... 28, 29

real time... 15, 16
relaxation............... 16, 38, 39, 41, 42, 49, 59, 62, 66, 70, 71, 78 82, 85, 87, 89, 95, 96, 98, 100, 105, 116
resonance............ 20, 24, 25, 26, 30, 31, 32, 33, 34, 37, 41, 77, 83 84, 85, 86, 101, 102, 104, 105, 106
resonate... 20
respiration.. 1, 11, 45, 48, 116
respiratory arterial pressure wave................17, 18, 26, 36, 38
respiratory sinus arrhythmia................... 11, 17, 21, 26, 34, 116
RSA.. 11, 21, 34

senescence.. 41
Shifu John Painter..27
sinewave... 29, 50, 73, 92, 93, 95
sinusoidal model... 14, 15
skeletal muscle... 2, 6, 79
sleep dysfunction... 44, 74
sleeping problems.. 7
SMR (low beta)... 108
somatic nervous system....................................... 19, 81, 96
somatic subsystem... 2
spectral analysis... 21, 23
sport... 26, 42, 106, 113
stillness... 32, 38, 39, 85, 87, 89
stress........... 10, 11, 12, 37, 45, 46, 47, 50, 51, 53, 54, 57, 63, 64
65, 67, 69, 70, 71, 78, 98, 100, 112, 116, 124, 126
subconscious.. 103
subjective time quanta...................................... 40, 41
suksma-gati.. 88
survival.. 3, 6, 8, 11
sympathetic..... 2, 4, 5, 6, 7, 8, 10, 11, 12, 14, 15, 16, 17, 18, 22, 23
30, 31, 32, 34, 35, 36, 37, 47, 75, 78, 79, 80, 81, 82, 83, 88, 89, 95
101, 102, 104, 105, 106, 109
sympathetic dominance....................................... 7, 8, 75, 81, 102
sympathetic emphasis..... 4, 5, 7, 8, 11, 12, 15, 16, 17, 18, 22, 31
34, 35, 36, 47, 79, 81, 83, 89, 95, 102, 104, 106, 109
sympathetic withdrawal....................................... 4, 5, 109

taijiquan... 81, 82
Taoist... 42, 81, 82, 83, 114
temporo-mandibular joint dysfunction.................................... 7
Test of Variables of Attention (TOVA)... 46, 47, 59, 61, 72, 109
thermal feedback... 109
thermal response.. 46
thoracic pressure..17
throat..3
tongue..3
traumatic brain injury................................ 44, 46, 56, 74, 119

unity 27, 29
urethral sphincter 3
urinary 3

valley 10, 14, 15, 16, 26, 35, 36, 41, 92, 105, 106
Van Der Pol 20
venous reservoirs 24
vertebrate physiology 10, 20

wholeness 27, 31, 102

yang 81, 95
yin 81, 95
yoga 33, 42, 77, 78, 80, 81, 83, 85, 86, 89, 106, 110, 111, 113, 114, 115, 116, 119
yogic theory 28
yoni mudra 86, 87

Zen 87, 88, 111, 114

About the Authors

Stephen Elliott

Stephen Elliott is a long-term student, practitioner, and teacher of Eastern yogic and martial arts and an avid life sciences researcher. He possesses a unique "systems view" resulting from a synthesis of diverse fields of knowledge including physiology, engineering, esoteric arts, and alternative medicine, as well as a deep understanding of yoga and meditation that can only be forged via direct experience. Stephen is a prolific inventor with over 30 patents granted or pending in areas of life sciences and telecommunications systems. He is Founder and President of COHERENCE® and is recognized in Who's Who of Executives and Professionals. Stephen continues to research the psycho-physiology of breathing relative to health, longevity, and esoteric experience, as well as actively investigate the relationship of suboptimal breathing to today's pandemic health challenges.

Dee Edmonson

"Integrative" neurotherapist Dee Edmonson, RN, Fellow BCIAC-EEG, is the Director of Neurotherapy services at a center in Plano, Texas. Specializing in the treatment of traumatic brain injury, attention disorder, addiction, depression, and stress/anxiety, she works with clients on a partnership basis incorporating conventional and alternative methods to maximize the body/mind's natural healing potential.

Dee has extensive experience in the fields of cardiology, neurology, and psychiatry, and a lifelong interest in the psychophysiology of breath. Her career experience includes developing innovative clinical programs for hospitals, consulting for medical equipment manufacturers, and serving as health and well-being educator in the areas of neurology and heart disease. She has been a recipient of the American Heart Association – Leadership Award, Southwest Dallas County. Dee also maintains a private practice – Neurologics^{SM} – and is a certified HeartMath 1:1 Provider.

Both Stephen and Dee offer workshops and seminars in Coherent Breathing. For details of these, visit: www.coherence.com and www.neurologics.us, respectively.

The New Science of Breath 42510